Statistical Tables

Fourth Edition

For students of

Science
Engineering
Psychology
Business
Management
Finance

J. Murdoch and J.A Barnes

palgrave
macmillan

Preface to the Fourth Edition

Much has changed in statistical analysis since these tables were first published and indeed since they were last revised. The biggest change has been the development and wide availability of personal computers together with comprehensive software which has become steadily easier to use. Between them, they have automated much that once had to be calculated manually as well as making possible previously impracticable methods of analysis.

However, those learning the subject should still find value in a set of tables such as these. The understanding of statistical concepts and the calculations which support them comes from working through practice problems, ideally with a capable teacher to provide assistance. Part of this learning process is finding out how to use tables – knowing what is tabulated and why and thus how to access the relevant table and how to interpret the result when it has been found. An important feature of some of the tables is the encouragement to consider the use of approximations – something which is basic to the application of statistical models to the real world. The tables should also be useful to the practitioner on those occasions where it is not convenient to have access to a computer.

Tables have been added, principally for distribution-free methods and for control chart applications. Some others – the basic mathematical tables of such as logarithms, squares and square roots – have been left out as their function is duplicated on readily available electronic calculators. As before, examples of the use of some of the tables have been given.

Permissions from other copyright holders are acknowledged with thanks at the foot of the relevant tables. Every effort has been made to trace all the copyright holders but if any have been inadvertently overlooked the publishers will be pleased to make the necessary arrangements at the first opportunity.

Comments on these tables and suggestions for their amendment will be welcome. Please either write to John Barnes care of the publishers or contact him by e-mail at J.A.Barnes@cranfield.ac.uk

Cranfield 1998

J. Murdoch
J. A. Barnes

First edition 1968
Second edition 1970
Third edition 1986
Fourth edition 1998

Published by
PALGRAVE MACMILLAN
Houndmills, Basingstoke, Hampshire RG21 6XS and
175 Fifth Avenue, New York, N.Y. 10010
Companies and representatives throughout the world

PALGRAVE MACMILLAN is the global academic imprint of the Palgrave Macmillan division of St. Martin's Press, LLC and of Palgrave Macmillan Ltd. Macmillan® is a registered trademark in the United States, United Kingdom and other countries. Palgrave is a registered trademark in the European Union and other countries.

ISBN–13: 978–0–333–55859–1
ISBN–10: 0–333–55859–6

This book is printed on paper suitable for recycling and made from fullymanaged and sustained forest sources. Logging, pulping and manufacturing processes are expected to conform to the environmental regulations of the country of origin.

A catalogue record for this book is available from the British Library.

14 13 12 11
11 10 09 08

Printed in China

Contents

Table   Basic Distribution Tables

1	Cumulative Binomial Probabilities	4
2	Cumulative Poisson Probabilities	8
3	Areas in Upper Tail of the Normal Distribution	13
4	Percentage Points of the Normal Distribution	14
5	Ordinates of the Normal Distribution	14
6	Exponential Function e^{-x}	15
7	Percentage Points of the t Distribution	17
8	Percentage Points of the χ^2 Distribution	18
9	Percentage Points of the F Distribution	20
10	Percentage Points of the Correlation Coefficient	22
11	Tukey's Wholly Significant Difference (5% Level)	23

Distribution-free (Non-parametric) Tables

12	Percentage Points of Spearman's Rank Correlation Coefficient	24
13	Percentage Points of Kendall's Rank Correlation Coefficient	25
14	Percentage Points of Nair's 'Studentised' Extreme Deviate from the Mean	26
15	Upper Percentage Points of Dixon's Rank Difference Ratio	27
16	Percentage Points of D in the One-sample Kolmogorov–Smirnov Distribution	28
17	Lower Percentage Points of the Wilcoxon Signed-rank Distribution	29
18	Percentage Points of D in the Two-sample Kolmogorov–Smirnov Distribution	30
19	Percentage Points of the Mann–Whitney U-Distribution	36
20	Percentage Points of Friedman's Distribution	38
21	Upper Tails of the Kruskal–Wallis Distribution	40

Statistical Process Control Tables

22	Control Chart Factors for Sample Mean ($\overline{X}$)	42
23	Control Chart Factors for Sample Range (using $\overline{R}$)	43
24	Control Chart Factors for Sample Range (using σ)	43
25	Control Chart Factors for Mean and Range (American usage)	44
26	Control Chart Factors for Standard Deviation (American usage)	45
27	Tolerance Factors for the Normal Distribution	46
28	Sample Size for Two-sided Distribution-free Tolerance Limits	47
29	Sample Size for One-sided Distribution-free Tolerance Limits	47

Critical Values for Runs

	Notes on Tables 30, 31, 32, 33, 34	48
30	Number of Runs on Either Side of the Mean: 5% Point	49
31	Number of Runs on Either Side of the Mean: 0.5% Point	49
32	Number of Runs Above and Below the Median	50
33	Lengths of Runs on Either Side of the Median: 5%, 1% and 0.1% Points	51
34	Critical Values of Lengths of Runs Up and Down	51

Attribute Single Sampling Tables

35	Derivation of Single Sampling Plans	52
36	Construction of O.C. Curves for Single Sampling Plans	53

Random Number Tables

37	Random Numbers	54
38	Random Standardised Normal Deviates (Z Values)	59

Financial Tables

39	Present Value Factors	60
40	Cumulative Present Value Factors	64
41	Capital Recovery Factors	68

Examples of the Use of Tables 11 to 16	72
Some Useful Formulae	77

Table 1 Cumulative Binomial Probabilities

p = probability of success in a single trial; n = number of trials. The table gives the probability of obtaining r *or more* successes in n independent trials. That is

$$\sum_{x=r}^{n} \binom{n}{x} p^x (1-p)^{n-x}$$

When there is no entry for a particular pair of values of r and p, this indicates that the appropriate probability is less than 0.000 05. Similarly, except for the case $r = 0$, when the entry is exact, a tabulated value of 1.0000 represents a probability greater than 0.999 95.

	$p =$	0.01	0.02	0.03	0.04	0.05	0.06	0.07	0.08	0.09
$n = 2$	$r = 0$	1.0000	1.0000	1.0000	1.0000	1.0000	1.0000	1.0000	1.0000	1.0000
	1	.0199	.0396	.0591	.0784	.0975	.1164	.1351	.1536	.1719
	2	.0001	.0004	.0009	.0016	.0025	.0036	.0049	.0064	.0081
$n = 5$	$r = 0$	1.0000	1.0000	1.0000	1.0000	1.0000	1.0000	1.0000	1.0000	1.0000
	1	.0490	.0961	.1413	.1846	.2262	.2661	.3043	.3409	.3760
	2	.0010	.0038	.0085	.0148	.0226	.0319	.0425	.0544	.0674
	3		.0001	.0003	.0006	.0012	.0020	.0031	.0045	.0063
	4						.0001	.0001	.0002	.0003
$n = 10$	$r = 0$	1.0000	1.0000	1.0000	1.0000	1.0000	1.0000	1.0000	1.0000	1.0000
	1	.0956	.1829	.2626	.3352	.4013	.4614	.5160	.5656	.6106
	2	.0043	.0162	.0345	.0582	.0861	.1176	.1517	.1879	.2254
	3	.0001	.0009	.0028	.0062	.0115	.0188	.0283	.0401	.0540
	4			.0001	.0004	.0010	.0020	.0036	.0058	.0088
	5					.0001	.0002	.0003	.0006	.0010
	6									.0001
$n = 20$	$r = 0$	1.0000	1.0000	1.0000	1.0000	1.0000	1.0000	1.0000	1.0000	1.0000
	1	.1821	.3324	.4562	.5580	.6415	.7099	.7658	.8113	.8484
	2	.0169	.0599	.1198	.1897	.2642	.3395	.4131	.4831	.5484
	3	.0010	.0071	.0210	.0439	.0755	.1150	.1610	.2121	.2666
	4		.0006	.0027	.0074	.0159	.0290	.0471	.0706	.0993
	5			.0003	.0010	.0026	.0056	.0107	.0183	.0290
	6				.0001	.0003	.0009	.0019	.0038	.0068
	7						.0001	.0003	.0006	.0013
	8								.0001	.0002
$n = 50$	$r = 0$	1.0000	1.0000	1.0000	1.0000	1.0000	1.0000	1.0000	1.0000	1.0000
	1	.3950	.6358	.7819	.8701	.9231	.9547	.9734	.9845	.9910
	2	.0894	.2642	.4447	.5995	.7206	.8100	.8735	.9173	.9468
	3	.0138	.0784	.1892	.3233	.4595	.5838	.6892	.7740	.8395
	4	.0016	.0178	.0628	.1391	.2396	.3527	.4673	.5747	.6697
	5	.0001	.0032	.0168	.0490	.1036	.1794	.2710	.3710	.4723
	6		.0005	.0037	.0144	.0378	.0776	.1350	.2081	.2928
	7		.0001	.0007	.0036	.0118	.0289	.0583	.1019	.1596
	8			.0001	.0008	.0032	.0094	.0220	.0438	.0768
	9				.0001	.0008	.0027	.0073	.0167	.0328
	10					.0002	.0007	.0022	.0056	.0125
	11						.0002	.0006	.0017	.0043
	12							.0001	.0005	.0013
	13								.0001	.0004
	14									.0001

Table 1 Cumulative Binomial Probabilities – continued

	p =	0.01	0.02	0.03	0.04	0.05	0.06	0.07	0.08	0.09
n = 100	r = 0	1.0000	1.0000	1.0000	1.0000	1.0000	1.0000	1.0000	1.0000	1.0000
	1	.6340	.8674	.9524	.9831	.9941	.9979	.9993	.9998	.9999
	2	.2642	.5967	.8054	.9128	.9629	.9848	.9940	.9977	.9991
	3	.0794	.3233	.5802	.7679	.8817	.9434	.9742	.9887	.9952
	4	.0184	.1410	.3528	.5705	.7422	.8570	.9256	.9633	.9827
	5	.0034	.0508	.1821	.3711	.5640	.7232	.8368	.9097	.9526
	6	.0005	.0155	.0808	.2116	.3840	.5593	.7086	.8201	.8955
	7	.0001	.0041	.0312	.1064	.2340	.3936	.5557	.6968	.8060
	8		.0009	.0106	.0475	.1280	.2517	.4012	.5529	.6872
	9		.0002	.0032	.0190	.0631	.1463	.2660	.4074	.5506
	10			.0009	.0068	.0282	.0775	.1620	.2780	.4125
	11			.0002	.0022	.0115	.0376	.0908	.1757	.2882
	12				.0007	.0043	.0168	.0469	.1028	.1876
	13				.0002	.0015	.0069	.0224	.0559	.1138
	14					.0005	.0026	.0099	.0282	.0645
	15					.0001	.0009	.0041	.0133	.0341
	16						.0003	.0016	.0058	.0169
	17						.0001	.0006	.0024	.0078
	18							.0002	.0009	.0034
	19							.0001	.0003	.0014
	20								.0001	.0005
	21									.0002
	22									.0001

	p =	0.10	0.15	0.20	0.25	0.30	0.35	0.40	0.45	0.50
n = 2	r = 0	1.0000	1.0000	1.0000	1.0000	1.0000	1.0000	1.0000	1.0000	1.0000
	1	.1900	.2775	.3600	.4375	.5100	.5775	.6400	.6975	.7500
	2	.0100	.0225	.0400	.0625	.0900	.1225	.1600	.2025	.2500
n = 5	r = 0	1.0000	1.0000	1.0000	1.0000	1.0000	1.0000	1.0000	1.0000	1.0000
	1	.4095	.5563	.6723	.7627	.8319	.8840	.9222	.9497	.9688
	2	.0815	.1648	.2627	.3672	.4718	.5716	.6630	.7438	.8125
	3	.0086	.0266	.0579	.1035	.1631	.2352	.3174	.4069	.5000
	4	.0005	.0022	.0067	.0156	.0308	.0540	.0870	.1312	.1875
	5		.0001	.0003	.0010	.0024	.0053	.0102	.0185	.0313
n = 10	r = 0	1.0000	1.0000	1.0000	1.0000	1.0000	1.0000	1.0000	1.0000	1.0000
	1	.6513	.8031	.8926	.9437	.9718	.9865	.9940	.9975	.9990
	2	.2639	.4557	.6242	.7560	.8507	.9140	.9536	.9767	.9893
	3	.0702	.1798	.3222	.4744	.6172	.7384	.8327	.9004	.9453
	4	.0128	.0500	.1209	.2241	.3504	.4862	.6177	.7430	.8281
	5	.0016	.0099	.0328	.0781	.1503	.2485	.3669	.4956	.6230
	6	.0001	.0014	.0064	.0197	.0473	.0949	.1662	.2616	.3770
	7		.0001	.0009	.0035	.0106	.0260	.0548	.1020	.1719
	8			.0001	.0004	.0016	.0048	.0123	.0274	.0547
	9					.0001	.0005	.0017	.0045	.0107
	10							.0001	.0003	.0010
n = 20	r = 0	1.0000	1.0000	1.0000	1.0000	1.0000	1.0000	1.0000	1.0000	1.0000
	1	.8784	.9612	.9885	.9968	.9992	.9998	1.0000	1.0000	1.0000
	2	.6083	.8244	.9308	.9757	.9924	.9979	.9995	.9999	1.0000
	3	.3231	.5951	.7939	.9087	.9645	.9879	.9964	.9991	.9998
	4	.1330	.3523	.5886	.7748	.8929	.9556	.9840	.9951	.9987
	5	.0432	.1702	.3704	.5852	.7625	.8818	.9490	.9811	.9941
	6	.0113	.0673	.1958	.3828	.5836	.7546	.8744	.9447	.9793
	7	.0024	.0219	.0867	.2142	.3920	.5834	.7500	.8701	.9423
	8	.0004	.0059	.0321	.1018	.2277	.3990	.5841	.7480	.8684
	9	.0001	.0013	.0100	.0409	.1133	.2376	.4044	.5857	.7483
	10		.0002	.0026	.0139	.0480	.1218	.2447	.4086	.5881
	11			.0006	.0039	.0171	.0532	.1275	.2493	.4119
	12			.0001	.0009	.0051	.0196	.0565	.1308	.2517
	13				.0002	.0013	.0060	.0210	.0580	.1316
	14					.0003	.0015	.0065	.0214	.0577
	15						.0003	.0016	.0064	.0207
	16							.0003	.0015	.0059
	17								.0003	.0013
	18									.0002

Table 1 Cumulative Binomial Probabilites – continued

	$p =$	0.10	0.15	0.20	0.25	0.30	0.35	0.40	0.45	0.50
$n = 50$	$r = 0$	1.0000	1.0000	1.0000	1.0000	1.0000	1.0000	1.0000	1.0000	1.0000
	1	.9948	.9997	1.0000	1.0000	1.0000	1.0000	1.0000	1.0000	1.0000
	2	.9662	.9971	.9998	1.0000	1.0000	1.0000	1.0000	1.0000	1.0000
	3	.8883	.9858	.9987	.9999	1.0000	1.0000	1.0000	1.0000	1.0000
	4	.7497	.9540	.9943	.9995	1.0000	1.0000	1.0000	1.0000	1.0000
	5	.5688	.8879	.9815	.9979	.9998	1.0000	1.0000	1.0000	1.0000
	6	.3839	.7806	.9520	.9930	.9993	.9999	1.0000	1.0000	1.0000
	7	.2298	.6387	.8966	.9806	.9975	.9998	1.0000	1.0000	1.0000
	8	.1221	.4812	.8096	.9547	.9927	.9992	.9999	1.0000	1.0000
	9	.0579	.3319	.6927	.9084	.9817	.9975	.9998	1.0000	1.0000
	10	.0245	.2089	.5563	.8363	.9598	.9933	.9992	.9999	1.0000
	11	.0094	.1199	.4164	.7378	.9211	.9840	.9978	.9998	1.0000
	12	.0032	.0628	.2893	.6184	.8610	.9658	.9943	.9994	1.0000
	13	.0010	.0301	.1861	.4890	.7771	.9339	.9867	.9982	.9998
	14	.0003	.0132	.1106	.3630	.6721	.8837	.9720	.9955	.9995
	15	.0001	.0053	.0607	.2519	.5532	.8122	.9460	.9896	.9987
	16		.0019	.0308	.1631	.4308	.7199	.9045	.9780	.9967
	17		.0007	.0144	.0983	.3161	.6111	.8439	.9573	.9923
	18		.0002	.0063	.0551	.2178	.4940	.7631	.9235	.9836
	19		.0001	.0025	.0287	.1406	.3784	.6644	.8727	.9675
	20			.0009	.0139	.0848	.2736	.5535	.8026	.9405
	21			.0003	.0063	.0478	.1861	.4390	.7138	.8987
	22			.0001	.0026	.0251	.1187	.3299	.6100	.8389
	23				.0010	.0123	.0710	.2340	.4981	.7601
	24				.0004	.0056	.0396	.1562	.3866	.6641
	25				.0001	.0024	.0207	.0978	.2840	.5561
	26					.0009	.0100	.0573	.1966	.4439
	27					.0003	.0045	.0314	.1279	.3359
	28					.0001	.0019	.0160	.0780	.2399
	29						.0007	.0076	.0444	.1611
	30						.0003	.0034	.0235	.1013
	31						.0001	.0014	.0116	.0595
	32							.0005	.0053	.0325
	33							.0002	.0022	.0164
	34							.0001	.0009	.0077
	35								.0003	.0033
	36								.0001	.0013
	37									.0005
	38									.0002

Table 1 gives binomial probabilities only for a limited range of values of n and p since, in practice, either the more compact tabulation of the Poisson distribution (Table 2) or that of the Normal distribution (Table 3) can usually be used to give an adequate approximation.

As a reasonable working rule:

(i) use the Poisson approximation if $p < 0.1$, putting $m = np$

(ii) use the Normal approximation if $0.1 \leq p \leq 0.9$ and $np > 5$, putting $\mu = np$ and $\sigma = \sqrt{np(1 - p)}$.

(iii) use the Poisson approximation if $p > 0.9$, putting $m = n(1 - p)$ and working in terms of 'failures'.

Note: For values of $p > 0.5$, work in terms of 'failures' which will have probability q ($= 1 - p$).

Example: What is the probability that 40 or more seeds will germinate out of 50 if the germination rate is 70%? Since the probability of 'success' is greater than 0.5, the table can not be used directly; however, 40 or more successes is the same as 10 or fewer 'failures'. The probability of 10 or fewer 'failures' = 1 – probability of 11 or more 'failures' = $1 - 0.9211 = 0.0789$.

Table 1 Cumulative Binomial Probabilities – continued

		p = 0.10	0.15	0.20	0.25	0.30	0.35	0.40	0.45	0.50
n = 100	r = 0	1.0000	1.0000	1.0000	1.0000	1.0000	1.0000	1.0000	1.0000	1.0000
	1	1.0000	1.0000	1.0000	1.0000	1.0000	1.0000	1.0000	1.0000	1.0000
	2	.9997	1.0000	1.0000	1.0000	1.0000	1.0000	1.0000	1.0000	1.0000
	3	.9981	1.0000	1.0000	1.0000	1.0000	1.0000	1.0000	1.0000	1.0000
	4	.9922	.9999	1.0000	1.0000	1.0000	1.0000	1.0000	1.0000	1.0000
	5	.9763	.9996	1.0000	1.0000	1.0000	1.0000	1.0000	1.0000	1.0000
	6	.9424	.9984	1.0000	1.0000	1.0000	1.0000	1.0000	1.0000	1.0000
	7	.8828	.9953	.9999	1.0000	1.0000	1.0000	1.0000	1.0000	1.0000
	8	.7939	.9878	.9997	1.0000	1.0000	1.0000	1.0000	1.0000	1.0000
	9	.6791	.9725	.9991	1.0000	1.0000	1.0000	1.0000	1.0000	1.0000
	10	.5487	.9449	.9977	1.0000	1.0000	1.0000	1.0000	1.0000	1.0000
	11	.4168	.9006	.9943	.9999	1.0000	1.0000	1.0000	1.0000	1.0000
	12	.2970	.8365	.9874	.9996	1.0000	1.0000	1.0000	1.0000	1.0000
	13	.1982	.7527	.9747	.9990	1.0000	1.0000	1.0000	1.0000	1.0000
	14	.1239	.6526	.9531	.9975	.9999	1.0000	1.0000	1.0000	1.0000
	15	.0726	.5428	.9196	.9946	.9998	1.0000	1.0000	1.0000	1.0000
	16	.0399	.4317	.8715	.9889	.9996	1.0000	1.0000	1.0000	1.0000
	17	.0206	.3275	.8077	.9789	.9990	1.0000	1.0000	1.0000	1.0000
	18	.0100	.2367	.7288	.9624	.9978	.9999	1.0000	1.0000	1.0000
	19	.0046	.1628	.6379	.9370	.9955	.9999	1.0000	1.0000	1.0000
	20	.0020	.1065	.5398	.9005	.9911	.9997	1.0000	1.0000	1.0000
	21	.0008	.0663	.4405	.8512	.9835	.9992	1.0000	1.0000	1.0000
	22	.0003	.0393	.3460	.7886	.9712	.9983	1.0000	1.0000	1.0000
	23	.0001	.0221	.2611	.7136	.9521	.9966	.9999	1.0000	1.0000
	24		.0119	.1891	.6289	.9245	.9934	.9997	1.0000	1.0000
	25		.0061	.1314	.5383	.8864	.9879	.9994	1.0000	1.0000
	26		.0030	.0875	.4465	.8369	.9789	.9988	1.0000	1.0000
	27		.0014	.0558	.3583	.7756	.9649	.9976	.9999	1.0000
	28		.0006	.0342	.2776	.7036	.9442	.9954	.9998	1.0000
	29		.0003	.0200	.2075	.6232	.9152	.9916	.9996	1.0000
	30		.0001	.0112	.1495	.5377	.8764	.9852	.9992	1.0000
	31			.0061	.1038	.4509	.8270	.9752	.9985	1.0000
	32			.0031	.0693	.3669	.7669	.9602	.9970	.9999
	33			.0016	.0446	.2893	.6971	.9385	.9945	.9998
	34			.0007	.0276	.2207	.6197	.9087	.9902	.9996
	35			.0003	.0164	.1629	.5376	.8697	.9834	.9991
	36			.0001	.0094	.1161	.4542	.8205	.9728	.9982
	37			.0001	.0052	.0799	.3731	.7614	.9571	.9967
	38				.0027	.0530	.2976	.6932	.9349	.9940
	39				.0014	.0340	.2301	.6178	.9049	.9895
	40				.0007	.0210	.1724	.5379	.8657	.9824
	41				.0003	.0125	.1250	.4567	.8169	.9716
	42				.0001	.0072	.0877	.3775	.7585	.9557
	43				.0001	.0040	.0594	.3033	.6913	.9334
	44					.0021	.0389	.2365	.6172	.9033
	45					.0011	.0246	.1789	.5387	.8644
	46					.0005	.0150	.1311	.4587	.8159
	47					.0003	.0088	.0930	.3804	.7579
	48					.0001	.0050	.0638	.3069	.6914
	49					.0001	.0027	.0423	.2404	.6178
	50						.0015	.0271	.1827	.5398
	51						.0007	.0168	.1346	.4602
	52						.0004	.0100	.0960	.3822
	53						.0002	.0058	.0662	.3086
	54						.0001	.0032	.0441	.2421
	55							.0017	.0284	.1841
	56							.0009	.0176	.1356
	57							.0004	.0106	.0967
	58							.0002	.0061	.0666
	59							.0001	.0034	.0443
	60								.0018	.0284
	61								.0009	.0176
	62								.0005	.0105
	63								.0002	.0060
	64								.0001	.0033
	65									.0018
	66									.0009
	67									.0004
	68									.0002
	69									.0001

Table 2 Cumulative Poisson Probabilities

The table gives the probability that *r or more* random events are contained in an interval when the average number of such events per interval is *m*, i.e.

$$\sum_{x=r}^{\infty} e^{-m}\frac{m^x}{x!}$$

Where there is no entry for a particular pair of values of *r* and *m*, this indicates that the appropriate probability is less than 0.000 05. Similarly, except for the case *r* = 0 when the entry is exact, a tabulated value of 1.0000 represents a probability greater than 0.999 95.

m =	0.1	0.2	0.3	0.4	0.5	0.6	0.7	0.8	0.9	1.0
r = 0	1.0000	1.0000	1.0000	1.0000	1.0000	1.0000	1.0000	1.0000	1.0000	1.0000
1	.0952	.1813	.2592	.3297	.3935	.4512	.5034	.5507	.5934	.6321
2	.0047	.0175	.0369	.0616	.0902	.1219	.1558	.1912	.2275	.2642
3	.0002	.0011	.0036	.0079	.0144	.0231	.0341	.0474	.0629	.0803
4		.0001	.0003	.0008	.0018	.0034	.0058	.0091	.0135	.0190
5				.0001	.0002	.0004	.0008	.0014	.0023	.0037
6							.0001	.0002	.0003	.0006
7										.0001

m =	1.1	1.2	1.3	1.4	1.5	1.6	1.7	1.8	1.9	2.0
r = 0	1.0000	1.0000	1.0000	1.0000	1.0000	1.0000	1.0000	1.0000	1.0000	1.0000
1	.6671	.6988	.7275	.7534	.7769	.7981	.8173	.8347	.8504	.8647
2	.3010	.3374	.3732	.4082	.4422	.4751	.5068	.5372	.5663	.5940
3	.0996	.1205	.1429	.1665	.1912	.2166	.2428	.2694	.2963	.3233
4	.0257	.0338	.0431	.0537	.0656	.0788	.0932	.1087	.1253	.1429
5	.0054	.0077	.0107	.0143	.0186	.0237	.0296	.0364	.0441	.0527
6	.0010	.0015	.0022	.0032	.0045	.0060	.0080	.0104	0132	.0166
7	.0001	.0003	.0004	.0006	.0009	.0013	.0019	.0026	.0034	.0045
8			.0001	.0001	.0002	.0003	.0004	.0006	.0008	.0011
9							.0001	.0001	.0002	.0002

m =	2.1	2.2	2.3	2.4	2.5	2.6	2.7	2.8	2.9	3.0
r = 0	1.0000	1.0000	1.0000	1.0000	1.0000	1.0000	1.0000	1.0000	1.0000	1.0000
1	.8775	.8892	.8997	.9093	.9179	.9257	.9328	.9392	.9450	.9502
2	.6204	.6454	.6691	.6916	.7127	.7326	.7513	.7689	.7854	.8009
3	.3504	.3773	.4040	.4303	.4562	.4816	.5064	.5305	.5540	.5768
4	.1614	.1806	.2007	.2213	.2424	.2640	.2859	.3081	.3304	.3528
5	.0621	.0725	.0838	.0959	.1088	.1226	.1371	.1523	.1682	.1847
6	.0204	.0249	.0300	.0357	.0420	.0490	.0567	.0651	.0742	.0839
7	.0059	.0075	.0094	.0116	.0142	.0172	.0206	.0244	.0287	.0335
8	.0015	.0020	.0026	.0033	.0042	.0053	.0066	.0081	.0099	.0119
9	.0003	.0005	.0006	.0009	.0011	.0015	.0019	.0024	.0031	.0038
10	.0001	.0001	.0001	.0002	.0003	.0004	.0005	.0007	.0009	.0011
11					.0001	.0001	.0001	.0002	.0002	.0003
12									.0001	.0001

Table 2 Cumulative Poisson Probabilities – continued

m =	3.1	3.2	3.3	3.4	3.5	3.6	3.7	3.8	3.9	4.0
r = 0	1.0000	1.0000	1.0000	1.0000	1.0000	1.0000	1.0000	1.0000	1.0000	1.0000
1	.9550	.9592	.9631	.9666	.9698	.9727	.9753	.9776	.9798	.9817
2	.8153	.8288	.8414	.8532	.8641	.8743	.8838	.8926	.9008	.9084
3	.5988	.6201	.6406	.6603	.6792	.6973	.7146	.7311	.7469	.7619
4	.3752	.3975	.4197	.4416	.4634	.4848	.5058	.5265	.5468	.5665
5	.2018	.2194	.2374	.2558	.2746	.2936	.3128	.3322	.3516	.3712
6	.0943	.1054	.1171	.1295	.1424	.1559	.1699	.1844	.1994	.2149
7	.0388	.0446	.0510	.0579	.0653	.0733	.0818	.0909	.1005	.1107
8	.0142	.0168	.0198	.0231	.0267	.0308	.0352	.0401	.0454	.0511
9	.0047	.0057	.0069	.0083	.0099	.0117	.0137	.0160	.0185	.0214
10	.0014	.0018	.0022	.0027	.0033	.0040	.0048	.0058	.0069	.0081
11	.0004	.0005	.0006	.0008	.0010	.0013	.0016	.0019	.0023	.0028
12	.0001	.0001	.0002	.0002	.0003	.0004	.0005	.0006	.0007	.0009
13				.0001	.0001	.0001	.0001	.0002	.0002	.0003
14									.0001	.0001

m =	4.1	4.2	4.3	4.4	4.5	4.6	4.7	4.8	4.9	5.0
r = 0	1.0000	1.0000	1.0000	1.0000	1.0000	1.0000	1.0000	1.0000	1.0000	1.0000
1	.9834	.9850	.9864	.9877	.9889	.9899	.9909	.9918	.9926	.9933
2	.9155	.9220	.9281	.9337	.9389	.9437	.9482	.9523	.9561	.9596
3	.7762	.7898	.8026	.8149	.8264	.8374	.8477	.8575	.8667	.8753
4	.5858	.6046	.6228	.6406	.6577	.6743	.6903	.7058	.7207	.7350
5	.3907	.4102	.4296	.4488	.4679	.4868	.5054	.5237	.5418	.5595
6	.2307	.2469	.2633	.2801	.2971	.3142	.3316	.3490	.3665	.3840
7	.1214	.1325	.1442	.1564	.1689	.1820	.1954	.2092	.2233	.2378
8	.0573	.0639	.0710	.0786	.0866	.0951	.1040	.1133	.1231	.1334
9	.0245	.0279	.0317	.0358	.0403	.0451	.0503	.0558	.0618	.0681
10	.0095	.0111	.0129	.0149	.0171	.0195	.0222	.0251	.0283	.0318
11	.0034	.0041	.0048	.0057	.0067	.0078	.0090	.0104	.0120	.0137
12	.0011	.0014	.0017	.0020	.0024	.0029	.0034	.0040	.0047	.0055
13	.0003	.0004	.0005	.0007	.0008	.0010	.0012	.0014	.0017	.0020
14	.0001	.0001	.0002	.0002	.0003	.0003	.0004	.0005	.0006	.0007
15				.0001	.0001	.0001	.0001	.0001	.0002	.0002
16									.0001	.0001

m =	5.2	5.4	5.6	5.8	6.0	6.2	6.4	6.6	6.8	7.0
r = 0	1.0000	1.0000	1.0000	1.0000	1.0000	1.0000	1.0000	1.0000	1.0000	1.0000
1	.9945	.9955	.9963	.9970	.9975	.9980	.9983	.9986	.9989	.9991
2	.9658	.9711	.9756	.9794	.9826	.9854	.9877	.9897	.9913	.9927
3	.8912	.9052	.9176	.9285	.9380	.9464	.9537	.9600	.9656	.9704
4	.7619	.7867	.8094	.8300	.8488	.8658	.8811	.8948	.9072	.9182
5	.5939	.6267	.6579	.6873	.7149	.7408	.7649	.7873	.8080	.8270
6	.4191	.4539	.4881	.5217	.5543	.5859	.6163	.6453	.6730	.6993
7	.2676	.2983	.3297	.3616	.3937	.4258	.4577	.4892	.5201	.5503
8	.1551	.1783	.2030	.2290	.2560	.2840	.3127	.3419	.3715	.4013
9	.0819	.0974	.1143	.1328	.1528	.1741	.1967	.2204	.2452	.2709
10	.0397	.0488	.0591	.0708	.0839	.0984	.1142	.1314	.1498	.1695
11	.0177	.0225	.0282	.0349	.0426	.0514	.0614	.0726	.0849	.0985
12	.0073	.0096	.0125	.0160	.0201	.0250	.0307	.0373	.0448	.0534
13	.0028	.0038	.0051	.0068	.0088	.0113	.0143	.0179	.0221	.0270
14	.0010	.0014	.0020	.0027	.0036	.0048	.0063	.0080	.0102	.0128
15	.0003	.0005	.0007	.0010	.0014	.0019	.0026	.0034	.0044	.0057
16	.0001	.0002	.0002	.0004	.0005	.0007	.0010	.0014	.0018	.0024
17		.0001	.0001	.0001	.0002	.0003	.0004	.0005	.0007	.0010
18					.0001	.0001	.0001	.0002	.0003	.0004
19								.0001	.0001	.0001

Table 2 Cumulative Poisson Probabilities – continued

m =	7.2	7.4	7.6	7.8	8.0	8.2	8.4	8.6	8.8	9.0
r = 0	1.0000	1.0000	1.0000	1.0000	1.0000	1.0000	1.0000	1.0000	1.0000	1.0000
1	.9993	.9994	.9995	.9996	.9997	.9997	.9998	.9998	.9998	.9999
2	.9939	.9949	.9957	.9964	.9970	.9975	.9979	.9982	.9985	.9988
3	.9745	.9781	.9812	.9839	.9862	.9882	.9900	.9914	.9927	.9938
4	.9281	.9368	.9446	.9515	.9576	.9630	.9677	.9719	.9756	.9788
5	.8445	.8605	.8751	.8883	.9004	.9113	.9211	.9299	.9379	.9450
6	.7241	.7474	.7693	.7897	.8088	.8264	.8427	.8578	.8716	.8843
7	.5796	.6080	.6354	.6616	.6866	.7104	.7330	.7543	.7744	.7932
8	.4311	.4607	.4900	.5188	.5470	.5746	.6013	.6272	.6522	.6761
9	.2973	.3243	.3518	.3796	.4075	.4353	.4631	.4906	.5177	.5443
10	.1904	.2123	.2351	.2589	.2834	.3085	.3341	.3600	.3863	.4126
11	.1133	.1293	.1465	.1648	.1841	.2045	.2257	.2478	.2706	.2940
12	.0629	.0735	.0852	.0980	.1119	.1269	.1429	.1600	.1780	.1970
13	.0327	.0391	.0464	.0546	.0638	.0739	.0850	.0971	.1102	.1242
14	.0159	.0195	.0238	.0286	.0342	.0405	.0476	.0555	.0642	.0739
15	.0073	.0092	.0114	.0141	.0173	.0209	.0251	.0299	.0353	.0415
16	.0031	.0041	.0052	.0066	.0082	.0102	.0125	.0152	.0184	.0220
17	.0013	.0017	.0022	.0029	.0037	.0047	.0059	.0074	.0091	.0111
18	.0005	.0007	.0009	.0012	.0016	.0021	.0027	.0034	.0043	.0053
19	.0002	.0003	.0004	.0005	.0006	.0009	.0011	.0015	.0019	.0024
20	.0001	.0001	.0001	.0002	.0003	.0003	.0005	.0006	.0008	.0011
21				.0001	.0001	.0001	.0002	.0002	.0003	.0004
22							.0001	.0001	.0001	.0002
23										.0001

m =	9.2	9.4	9.6	9.8	10.0	11.0	12.0	13.0	14.0	15.0
r = 0	1.0000	1.0000	1.0000	1.0000	1.0000	1.0000	1.0000	1.0000	1.0000	1.0000
1	.9999	.9999	.9999	.9999	1.0000	1.0000	1.0000	1.0000	1.0000	1.0000
2	.9990	.9991	.9993	.9994	.9995	.9998	.9999	1.0000	1.0000	1.0000
3	.9947	.9955	.9962	.9967	.9972	.9988	.9995	.9998	.9999	1.0000
4	.9816	.9840	.9862	.9880	.9897	.9951	.9977	.9990	.9995	.9998
5	.9514	.9571	.9622	.9667	.9707	.9849	.9924	.9963	.9982	.9991
6	.8959	.9065	.9162	.9250	.9329	.9625	.9797	.9893	.9945	.9972
7	.8108	.8273	.8426	.8567	.8699	.9214	.9542	.9741	.9858	.9924
8	.6990	.7208	.7416	.7612	.7798	.8568	.9105	.9460	.9684	.9820
9	.5704	.5958	.6204	.6442	.6672	.7680	.8450	.9002	.9379	.9626
10	.4389	.4651	.4911	.5168	.5421	.6595	.7576	.8342	.8906	.9301
11	.3180	.3424	.3671	.3920	.4170	.5401	.6528	.7483	.8243	.8815
12	.2168	.2374	.2588	.2807	.3032	.4207	.5384	.6468	.7400	.8152
13	.1393	.1552	.1721	.1899	.2084	.3113	.4240	.5369	.6415	.7324
14	.0844	.0958	.1081	.1214	.1355	.2187	.3185	.4270	.5356	.6368
15	.0483	.0559	.0643	.0735	.0835	.1460	.2280	.3249	.4296	.5343
16	.0262	.0309	.0362	.0421	.0487	.0926	.1556	.2364	.3306	.4319
17	.0135	.0162	.0194	.0230	.0270	.0559	.1013	.1645	.2441	.3359
18	.0066	.0081	.0098	.0119	.0143	.0322	.0630	.1095	.1728	.2511
19	.0031	.0038	.0048	.0059	.0072	.0177	.0374	.0698	.1174	.1805
20	.0014	.0017	.0022	.0028	.0035	.0093	.0213	.0427	.0765	.1248
21	.0006	.0008	.0010	.0012	.0016	.0047	.0116	.0250	.0479	.0830
22	.0002	.0003	.0004	.0005	.0007	.0023	.0061	.0141	.0288	.0531
23	.0001	.0001	.0002	.0002	.0003	.0010	.0030	.0076	.0167	.0327
24			.0001	.0001	.0001	.0005	.0015	.0040	.0093	.0195
25						.0002	.0007	.0020	.0050	.0122
26						.0001	.0003	.0010	.0026	.0062
27							.0001	.0005	.0013	.0033
28							.0001	.0002	.0006	.0017
29								.0001	.0003	.0009
30									.0001	.0004
31									.0001	.0002
32										.0001

Table 2 Cumulative Poisson Probabilities – continued

m =	16.0	17.0	18.0	19.0	20.0	21.0	22.0	23.0	24.0	25.0
r = 0	1.0000	1.0000	1.0000	1.0000	1.0000	1.0000	1.0000	1.0000	1.0000	1.0000
1	1.0000	1.0000	1.0000	1.0000	1.0000	1.0000	1.0000	1.0000	1.0000	1.0000
2	1.0000	1.0000	1.0000	1.0000	1.0000	1.0000	1.0000	1.0000	1.0000	1.0000
3	1.0000	1.0000	1.0000	1.0000	1.0000	1.0000	1.0000	1.0000	1.0000	1.0000
4	.9999	1.0000	1.0000	1.0000	1.0000	1.0000	1.0000	1.0000	1.0000	1.0000
5	.9996	.9998	.9999	1.0000	1.0000	1.0000	1.0000	1.0000	1.0000	1.0000
6	.9986	.9993	.9997	.9998	.9999	1.0000	1.0000	1.0000	1.0000	1.0000
7	.9960	.9979	.9990	.9995	.9997	.9999	.9999	1.0000	1.0000	1.0000
8	.9900	.9946	.9971	.9985	.9992	.9996	.9998	.9999	1.0000	1.0000
9	.9780	.9874	.9929	.9961	.9979	.9989	.9994	.9997	.9998	.9999
10	.9567	.9739	.9846	.9911	.9950	.9972	.9985	.9992	.9996	.9998
11	.9226	.9509	.9696	.9817	.9892	.9937	.9965	.9980	.9989	.9994
12	.8730	.9153	.9451	.9653	.9786	.9871	.9924	.9956	.9975	.9986
13	.8069	.8650	.9083	.9394	.9610	.9755	.9849	.9909	.9946	.9969
14	.7255	.7991	.8574	.9016	.9339	.9566	.9722	.9826	.9893	.9935
15	.6325	.7192	.7919	.8503	.8951	.9284	.9523	.9689	.9802	.9876
16	.5333	.6285	.7133	.7852	.8435	.8889	.9231	.9480	.9656	.9777
17	.4340	.5323	.6249	.7080	.7789	.8371	.8830	.9179	.9437	.9623
18	.3407	.4360	.5314	.6216	.7030	.7730	.8310	.8772	.9129	.9395
19	.2577	.3450	.4378	.5305	.6186	.6983	.7675	.8252	.8717	.9080
20	.1878	.2637	.3491	.4394	.5297	.6157	.6940	.7623	.8197	.8664
21	.1318	.1945	.2693	.3528	.4409	.5290	.6131	.6899	.7574	.8145
22	.0892	.1385	.2009	.2745	.3563	.4423	.5284	.6106	.6861	.7527
23	.0582	.0953	.1449	.2069	.2794	.3595	.4436	.5277	.6083	.6825
24	.0367	.0633	.1011	.1510	.2125	.2840	.3626	.4449	.5272	.6061
25	.0223	.0406	.0683	.1067	.1568	.2178	.2883	.3654	.4460	.5266
26	.0131	.0252	.0446	.0731	.1122	.1623	.2229	.2923	.3681	.4471
27	.0075	.0152	.0282	.0486	.0779	.1174	.1676	.2277	.2962	.3706
28	.0041	.0088	.0173	.0313	.0525	.0825	.1225	.1726	.2323	.2998
29	.0022	.0050	.0103	.0195	.0343	.0564	.0871	.1274	.1775	.2366
30	.0011	.0027	.0059	.0118	.0218	.0374	.0602	.0915	.1321	.1821
31	.0006	.0014	.0033	.0070	.0135	.0242	.0405	.0640	.0958	.1367
32	.0003	.0007	.0018	.0040	.0081	.0152	.0265	.0436	.0678	.1001
33	.0001	.0004	.0010	.0022	.0047	.0093	.0169	.0289	.0467	.0715
34	.0001	.0002	.0005	.0012	.0027	.0055	.0105	.0187	.0314	.0498
35		.0001	.0002	.0006	.0015	.0032	.0064	.0118	.0206	.0338
36			.0001	.0003	.0008	.0018	.0038	.0073	.0132	.0225
37			.0001	.0002	.0004	.0010	.0022	.0044	.0082	.0146
38				.0001	.0002	.0005	.0012	.0026	.0050	.0092
39					.0001	.0003	.0007	.0015	.0030	.0057
40					.0001	.0001	.0004	.0008	.0017	.0034
41						.0001	.0002	.0004	.0010	.0020
42							.0001	.0002	.0005	.0012
43								.0001	.0003	.0007
44								.0001	.0002	.0004
45									.0001	.0002
46										.0001

Table 2 Cumulative Poisson Probabilities – continued

m =	26.0	27.0	28.0	29.0	30.0	32.0	34.0	36.0	38.0	40.0
r = 9	1.0000	1.0000	1.0000	1.0000	1.0000	1.0000	1.0000	1.0000	1.0000	1.0000
10	.9999	.9999	1.0000	1.0000	1.0000	1.0000	1.0000	1.0000	1.0000	1.0000
11	.9997	.9998	.9999	1.0000	1.0000	1.0000	1.0000	1.0000	1.0000	1.0000
12	.9992	.9996	.9998	.9999	.9999	1.0000	1.0000	1.0000	1.0000	1.0000
13	.9982	.9990	.9994	.9997	.9998	1.0000	1.0000	1.0000	1.0000	1.0000
14	.9962	.9978	.9987	.9993	.9996	.9999	1.0000	1.0000	1.0000	1.0000
15	.9924	.9954	.9973	.9984	.9991	.9997	.9999	1.0000	1.0000	1.0000
16	.9858	.9912	.9946	.9967	.9981	.9993	.9998	.9999	1.0000	1.0000
17	.9752	.9840	.9899	.9937	.9961	.9986	.9995	.9998	1.0000	1.0000
18	.9580	.9726	.9821	.9885	.9927	.9972	.9990	.9997	.9999	1.0000
19	.9354	.9555	.9700	.9801	.9871	.9948	.9980	.9993	.9998	.9999
20	.9032	.9313	.9522	.9674	.9781	.9907	.9963	.9986	.9995	.9998
21	.8613	.8985	.9273	.9489	.9647	.9841	.9932	.9973	.9990	.9996
22	.8095	.8564	.8940	.9233	.9456	.9740	.9884	.9951	.9981	.9993
23	.7483	.8048	.8517	.8896	.9194	.9594	.9809	.9915	.9965	.9986
24	.6791	.7441	.8002	.8471	.8854	.9390	.9698	.9859	.9938	.9974
25	.6041	.6758	.7401	.7958	.8428	.9119	.9540	.9776	.9897	.9955
26	.5261	.6021	.6728	.7363	.7916	.8772	.9326	.9655	.9834	.9924
27	.4481	.5256	.6003	.6699	.7327	.8344	.9047	.9487	.9741	.9877
28	.3730	.4491	.5251	.5986	.6671	.7838	.8694	.9264	.9611	.9807
29	.3033	.3753	.4500	.5247	.5969	.7259	.8267	.8977	.9435	.9706
30	.2407	.3065	.3774	.4508	.5243	.6620	.7765	.8621	.9204	.9568
31	.1866	.2447	.3097	.3794	.4516	.5939	.7196	.8194	.8911	.9383
32	.1411	.1908	.2485	.3126	.3814	.5235	.6573	.7697	.8552	.9145
33	.1042	.1454	.1949	.2521	.3155	.4532	.5911	.7139	.8125	.8847
34	.0751	.1082	.1495	.1989	.2556	.3850	.5228	.6530	.7635	.8486
35	.0528	.0787	.1121	.1535	.2027	.3208	.4546	.5885	.7086	.8061
36	.0363	.0559	.0822	.1159	.1574	.2621	.3883	.5222	.6490	.7576
37	.0244	.0388	.0589	.0856	.1196	.2099	.3256	.4558	.5862	.7037
38	.0160	.0263	.0413	.0619	.0890	.1648	.2681	.3913	.5216	.6453
39	.0103	.0175	.0283	.0438	.0648	.1268	.2166	.3301	.4570	.5840
40	.0064	.0113	.0190	.0303	.0463	.0956	.1717	.2737	.3941	.5210
41	.0039	.0072	.0125	.0205	.0323	.0707	.1336	.2229	.3343	.4581
42	.0024	.0045	.0080	.0136	.0221	.0512	.1019	.1783	.2789	.3967
43	.0014	.0027	.0050	.0089	.0148	.0364	.0763	.1401	.2288	.3382
44	.0008	.0016	.0031	.0056	.0097	.0253	.0561	.1081	.1845	.2838
45	.0004	.0009	.0019	.0035	.0063	.0173	.0404	.0819	.1462	.2343
46	.0002	.0005	.0011	.0022	.0040	.0116	.0286	.0609	.1139	.1903
47	.0001	.0003	.0006	.0013	.0025	.0076	.0199	.0445	.0872	.1521
48	.0001	.0002	.0004	.0008	.0015	.0049	.0136	.0320	.0657	.1196
49		.0001	.0002	.0004	.0009	.0031	.0091	.0225	.0486	.0925
50			.0001	.0002	.0005	.0019	.0060	.0156	.0353	.0703
51			.0001	.0001	.0003	.0012	.0039	.0106	.0253	.0526
52				.0001	.0002	.0007	.0024	.0071	.0178	.0387
53					.0001	.0004	.0015	.0047	.0123	.0281
54					.0001	.0002	.0009	.0030	.0084	.0200
55						.0001	.0006	.0019	.0056	.0140
56						.0001	.0003	.0012	.0037	.0097
57							.0002	.0007	.0024	.0066
58							.0001	.0005	.0015	.0044
59							.0001	.0003	.0010	.0029
60								.0002	.0006	.0019
61								.0001	.0004	.0012
62								.0001	.0002	.0008
63									.0001	.0005
64									.0001	.0003
65										.0002
66										.0001
67										.0001

For values of m greater than 30, use the table of areas under the Normal curve (Table 3) to obtain approximate Poisson probabilities, putting $\mu = m$ and $\sigma = \sqrt{m}$.

Table 3 Areas in Upper Tail of the Normal Distribution

The function tabulated is $1 - \Phi(z)$ where $\Phi(z)$ is the cumulative distribution function of a standardised Normal variable, z.

Thus $1 - \Phi(z) = \dfrac{1}{\sqrt{2\pi}} \displaystyle\int_{z}^{\infty} e^{-z^2/2}$ is the probability that a standardised Normal variate selected at random will be greater than a

value of $z \left(= \dfrac{x - \mu}{\sigma} \right)$

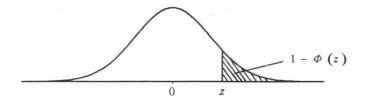

$\dfrac{x-\mu}{\sigma}$	.00	.01	.02	.03	.04	.05	.06	.07	.08	.09
0.0	.5000	.4960	.4920	.4880	.4840	.4801	.4761	.4721	.4681	.4641
0.1	.4602	.4562	.4522	.4483	.4443	.4404	.4364	.4325	.4286	.4247
0.2	.4207	.4168	.4129	.4090	.4052	.4013	.3974	.3936	.3897	.3859
0.3	.3821	.3783	.3745	.3707	.3669	.3632	.3594	.3557	.3520	.3483
0.4	.3446	.3409	.3372	.3336	.3300	.3264	.3228	.3192	.3156	.3121
0.5	.3085	.3050	.3015	.2981	.2946	.2912	.2877	.2843	.2810	.2776
0.6	.2743	.2709	.2676	.2643	.2611	.2578	.2546	.2514	.2483	.2451
0.7	.2420	.2389	.2358	.2327	.2296	.2266	.2236	.2206	.2177	.2148
0.8	.2119	.2090	.2061	.2033	.2005	.1977	.1949	.1922	.1894	.1867
0.9	.1841	.1814	.1788	.1762	.1736	.1711	.1685	.1660	.1635	.1611
1.0	.1587	.1562	.1539	.1515	.1492	.1469	.1446	.1423	.1401	.1379
1.1	.1357	.1335	.1314	.1292	.1271	.1251	.1230	.1210	.1190	.1170
1.2	.1151	.1131	.1112	.1093	.1075	.1056	.1038	.1020	.1003	.0985
1.3	.0968	.0951	.0934	.0918	.0901	.0885	.0869	.0853	.0838	.0823
1.4	.0808	.0793	.0778	.0764	.0749	.0735	.0721	.0708	.0694	.0681
1.5	.0668	.0655	.0643	.0630	.0618	.0606	.0594	.0582	.0571	.0559
1.6	.0548	.0537	.0526	.0516	.0505	.0495	.0485	.0475	.0465	.0455
1.7	.0446	.0436	.0427	.0418	.0409	.0401	.0392	.0384	.0375	.0367
1.8	.0359	.0351	.0344	.0336	.0329	.0322	.0314	.0307	.0301	.0294
1.9	.0287	.0281	.0274	.0268	.0262	.0256	.0250	.0244	.0239	.0233
2.0	.02275	.02222	.02169	.02118	.02068	.02018	.01970	.01923	.01876	.01831
2.1	.01786	.01743	.01700	.01659	.01618	.01578	.01539	.01500	.01463	.01426
2.2	.01390	.01355	.01321	.01287	.01255	.01222	.01191	.01160	.01130	.01101
2.3	.01072	.01044	.01017	.00990	.00964	.00939	.00914	.00889	.00866	.00842
2.4	.00820	.00798	.00776	.00755	.00734	.00714	.00695	.00676	.00657	.00639
2.5	.00621	.00604	.00587	.00570	.00554	.00539	.00523	.00508	.00494	.00480
2.6	.00466	.00453	.00440	.00427	.00415	.00402	.00391	.00379	.00368	.00357
2.7	.00347	.00336	.00326	.00317	.00307	.00298	.00289	.00280	.00272	.00264
2.8	.00256	.00248	.00240	.00233	.00226	.00219	.00212	.00205	.00199	.00193
2.9	.00187	.00181	.00175	.00169	.00164	.00159	.00154	.00149	.00144	.00139
3.0	.00135	.00131	.00126	.00122	.00118	.00114	.00111	.00107	.00104	.00100
3.1	.00097	.00094	.00090	.00087	.00084	.00082	.00079	.00076	.00074	.00071
3.2	.00069	.00066	.00064	.00062	.00060	.00058	.00056	.00054	.00052	.00050
3.3	.00048	.00047	.00045	.00043	.00042	.00040	.00039	.00038	.00036	.00035
3.4	.00034	.00032	.00031	.00030	.00029	.00028	.00027	.00026	.00025	.00024
3.5	.00023	.00022	.00022	.00021	.00020	.00019	.00019	.00018	.00017	.00017
3.6	.00016	.00015	.00015	.00014	.00014	.00013	.00013	.00012	.00012	.00011
3.7	.000108	.000104	.000100	.000096	.000092	.000088	.000085	.000082	.000078	.000075
3.8	.000072	.000069	.000067	.000064	.000062	.000059	.000057	.000054	.000052	.000050
3.9	.000048	.000046	.000044	.000042	.000041	.000039	.000037	.000036	.000034	.000033
4.0	.000032									

$5.0 \rightarrow 0.000\,000\,286\,7$ $5.5 \rightarrow 0.000\,000\,019\,0$ $6.0 \rightarrow 0.000\,000\,001\,0$

Table 4 Percentage Points of the Normal Distribution

The table gives the 100α percentage points, z_α of a standardised normal distribution where

$$\alpha = \frac{1}{\sqrt{2\pi}}\int_{z_\alpha}^{\infty} e^{-z^2/2}\,dz\,.$$

Thus z_α is the value of a standardised normal variate which has probability α of being exceeded.

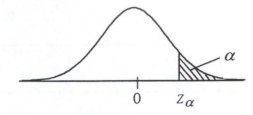

α	z_α	α	z_α	α	z_α	α	z_α	α	z_α	α	z_α
.50	0.0000	.050	1.6449	.030	1.8808	.020	2.0537	.010	2.3263	.050	1.6449
.45	0.1257	.048	1.6646	.029	1.8957	.019	2.0749	.009	2.3656	.010	2.3263
.40	0.2533	.046	1.6849	.028	1.9910	.018	2.0969	.008	2.4089	.001	3.0902
.35	0.3853	.044	1.7060	.027	1.9268	.017	2.1201	.007	2.4573	.000 1	3.7190
.30	0.5244	.042	1.7279	.026	1.9431	.016	2.1444	.006	2.5121	.000 01	4.2649
.25	0.6745	.040	1.7507	.025	1.9600	.015	2.1701	.005	2.5758	.025	1.9600
.20	0.8416	.038	1.7744	.024	1.9774	.014	2.1973	.004	2.6521	.005	2.5758
.15	1.0364	.036	1.7991	.023	1.9954	.013	2.2262	.003	2.7478	.000 5	3.2905
.10	1.2816	.034	1.8250	.022	2.0141	.012	2.2571	.002	2.8782	.000 05	3.8906
.05	1.6449	.032	1.8522	.021	2.0335	.011	2.2904	.001	3.0902	.000 005	4.4172

Table 5 Ordinates of the Normal Distribution

The table gives $\phi(z)$ for values of the standardised normal variate, z, in the interval 0.0 (0.1) 4.0 where

$$\phi(z) = \frac{1}{\sqrt{2\pi}} e^{-z^2/2}\,.$$

z	.0	.1	.2	.3	.4	.5	.6	.7	.8	.9
0.0	.3989	.3970	.3910	.3814	.3683	.3521	.3332	.3123	.2897	.2661
1.0	.2420	.2179	.1942	.1714	.1497	.1295	.1109	.0940	.0790	.0656
2.0	.0540	.0440	.0355	.0283	.0224	.0175	.0136	.0104	.0079	.0060
3.0	.0044	.0033	.0024	.0017	.0012	.0009	.0006	.0004	.0003	.0002
4.0	.0001									

Table 6 Exponential Function e^{-x}

For any negative exponential distribution, the tabulated function may be used to find the proportion of the distribution in excess of x times the mean. As an example, in random sampling of an exponential variate with a mean of 8, the probability that a single value will exceed 6 is 0.4724 since 6 is 0.75 times the distribution mean. Further, the 1% point of the distribution is 4.61 times the mean.

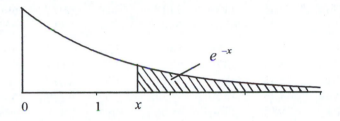

x	.0	.1	.2	.3	.4	.5	.6	.7	.8	.9
1.0	.3679	.3329	.3012	.2725	.2466	.2231	.2019	.1827	.1653	.1496
2.0	.1353	.1225	.1108	.1003	.0907	.0821	.0743	.0672	.0608	.0550
3.0	.0498	.0450	.0408	.0369	.0334	.0302	.0273	.0247	.0224	.0202
4.0	.0183	.0166	.0150	.0136	.0123	.0111	.0101	$.0^2910$	$.0^2823$	$.0^2745$
5.0	$.0^2674$	$.0^2610$	$.0^2552$	$.0^2499$	$.0^2452$	$.0^2409$	$.0^2370$	$.0^2335$	$.0^2303$	$.0^2274$
6.0	$.0^2248$	$.0^2224$	$.0^2203$	$.0^2184$	$.0^2166$	$.0^2150$	$.0^2136$	$.0^2123$	$.0^2111$	$.0^2101$
7.0	$.0^3912$	$.0^3825$	$.0^3747$	$.0^3676$	$.0^3611$	$.0^3553$	$.0^3500$	$.0^3453$	$.0^3410$	$.0^3371$
8.0	$.0^3335$	$.0^3304$	$.0^3275$	$.0^3249$	$.0^3225$	$.0^3203$	$.0^3184$	$.0^3167$	$.0^3151$	$.0^3136$
9.0	$.0^3123$	$.0^3112$	$.0^3101$	$.0^4914$	$.0^4827$	$.0^4749$	$.0^4677$	$.0^4613$	$.0^4555$	$.0^4502$
10.0	$.0^4454$	$.0^4411$	$.0^4372$	$.0^4336$	$.0^4304$	$.0^4275$	$.0^4249$	$.0^4225$	$.0^4204$	$.0^4185$
11.0	$.0^4167$	$.0^4151$	$.0^4137$	$.0^4124$	$.0^4112$	$.0^4101$	$.0^5917$	$.0^5829$	$.0^5750$	$.0^5679$
12.0	$.0^5614$	$.0^5556$	$.0^5502$	$.0^5455$	$.0^5412$	$.0^5373$	$.0^5337$	$.0^5305$	$.0^5276$	$.0^5250$
13.0	$.0^5226$	$.0^5205$	$.0^5185$	$.0^5167$	$.0^5152$	$.0^5137$	$.0^5124$	$.0^5112$	$.0^5102$	$.0^6919$
14.0	$.0^6832$	$.0^6752$	$.0^6681$	$.0^6616$	$.0^6557$	$.0^6504$	$.0^6456$	$.0^6413$	$.0^6374$	$.0^6338$
15.0	$.0^6306$	$.0^6277$	$.0^6250$	$.0^6227$	$.0^6205$	$.0^6186$	$.0^6168$	$.0^6152$	$.0^6137$	$.0^6124$
16.0	$.0^6113$	$.0^6102$	$.0^7921$	$.0^7834$	$.0^7754$	$.0^7683$	$.0^7618$	$.0^7559$	$.0^7506$	$.0^7458$
17.0	$.0^7414$	$.0^7375$	$.0^7339$	$.0^7307$	$.0^7278$	$.0^7251$	$.0^7227$	$.0^7206$	$.0^7186$	$.0^7168$
18.0	$.0^7152$	$.0^7138$	$.0^7125$	$.0^7113$	$.0^7102$	$.0^8924$	$.0^8836$	$.0^8756$	$.0^8684$	$.0^8619$
19.0	$.0^8560$	$.0^8507$	$.0^8459$	$.0^8415$	$.0^8376$	$.0^8340$	$.0^8307$	$.0^8278$	$.0^8252$	$.0^8228$
20.0	$.0^8206$									

To keep each entry compact, the superscript digit refers to the number of zeros which precede the first significant figure. Thus $0.0^3123 = 0.000\ 123$.

Table 6 Exponential Function e^{-x} – continued

x	.00	.01	.02	.03	.04	.05	.06	.07	.08	.09
0	1.0000	.9900	.9802	.9704	.9608	.9512	.9418	.9324	.9231	.9139
.1	.9048	.8958	.8869	.8781	.8694	.8607	.8521	.8437	.8353	.8270
.2	.8187	.8106	.8025	.7945	.7866	.7788	.7711	.7634	.7558	.7483
.3	.7408	.7334	.7261	.7189	.7118	.7047	.6977	.6907	.6839	.6771
.4	.6703	.6636	.6570	.6505	.6440	.6376	.6313	.6250	.6188	.6126
.5	.6065	.6005	.5945	.5886	.5827	.5770	.5712	.5655	.5599	.5543
.6	.5488	.5434	.5379	.5326	.5273	.5220	.5169	.5117	.5066	.5016
.7	.4966	.4916	.4868	.4819	.4771	.4724	.4677	.4630	.4584	.4538
.8	.4493	.4449	.4404	.4360	.4317	.4274	.4232	.4190	.4148	.4107
.9	.4066	.4025	.3985	.3946	.3906	.3867	.3829	.3791	.3753	.3716
1.0	.3679	.3642	.3606	.3570	.3535	.3499	.3465	.3430	.3396	.3362
1.1	.3329	.3296	.3263	.3230	.3198	.3166	.3135	.3104	.3073	.3042
1.2	.3012	.2892	.2952	.2923	.2894	.2865	.2837	.2808	.2780	.2753
1.3	.2725	.2698	.2671	.2645	.2618	.2592	.2567	.2541	.2516	.2491
1.4	.2466	.2441	.2417	.2393	.2369	.2346	.2322	.2299	.2276	.2254
1.5	.2231	.2209	.2187	.2165	.2144	.2122	.2101	.2080	.2060	.2039
1.6	.2019	.1999	.1979	.1959	.1940	.1920	.1901	.1882	.1864	.1845
1.7	.1827	.1809	.1791	.1773	.1755	.1738	.1720	.1703	.1686	.1670
1.8	.1653	.1637	.1620	.1604	.1588	.1572	.1557	.1541	.1526	.1511
1.9	.1496	.1481	.1466	.1451	.1437	.1423	.1409	.1395	.1381	.1367
2.0	.1353	.1340	.1327	.1313	.1300	.1287	.1275	.1262	.1249	.1237
2.1	.1225	.1212	.1200	.1188	.1177	.1165	.1153	.1142	.1130	.1119
2.2	.1108	.1097	.1086	.1075	.1065	.1054	.1044	.1035	.1023	.1013
2.3	.1003	.0993	.0983	.0973	.0963	.0954	.0944	.0935	.0926	.0916
2.4	.0907	.0898	.0889	.0880	.0872	.0863	.0854	.0846	.0837	.0829
2.5	.0821	.0813	.0805	.0797	.0789	.0781	.0773	.0765	.0758	.0750
2.6	.0743	.0735	.0728	.0721	.0714	.0707	.0699	.0693	.0686	.0679
2.7	.0672	.0665	.0659	.0652	.0646	.0639	.0633	.0627	.0620	.0614
2.8	.0608	.0602	.0596	.0590	.0584	.0578	.0573	.0567	.0561	.0556
2.9	.0550	.0545	.0539	.0534	.0529	.0523	.0518	.0513	.0508	.0503
3.0	.0498	.0493	.0488	.0483	.0478	.0474	.0469	.0464	.0460	.0455
3.1	.0450	.0446	.0442	.0437	.0433	.0429	.0424	.0420	.0416	.0412
3.2	.0408	.0404	.0400	.0396	.0392	.0388	.0384	.0380	.0376	.0373
3.3	.0369	.0365	.0362	.0358	.0354	.0351	.0347	.0344	.0340	.0337
3.4	.0334	.0330	.0327	.0324	.0321	.0317	.0314	.0311	.0308	.0305
3.5	.0302	.0299	.0296	.0293	.0290	.0287	.0284	.0282	.0279	.0276
3.6	.0273	.0271	.0268	.0265	.0263	.0260	.0257	.0255	.0252	.0250
3.7	.0247	.0245	.0242	.0240	.0238	.0235	.0233	.0231	.0228	.0226
3.8	.0224	.0221	.0219	.0217	.0215	.0213	.0211	.0209	.0207	.0204
3.9	.0202	.0200	.0198	.0196	.0194	.0193	.0191	.0189	.0187	.0185
4.0	.0183	.0181	.0180	.0178	.0176	.0174	.0172	.0171	.0169	.0167
4.1	.0166	.0164	.0162	.0161	.0159	.0158	.0156	.0155	.0153	.0151
4.2	.0150	.0148	.0147	.0146	.0144	.0143	.0141	.0140	.0138	.0137
4.3	.0136	.0134	.0133	.0132	.0130	.0129	.0128	.0127	.1025	.0124
4.4	.0123	.0122	.0120	.0119	.0118	.0117	.0116	.0114	.0113	.0112
4.5	.0111	.0110	.0109	.0108	.0107	.0106	.0105	.0104	.0103	.0102
4.6	.0101	.0100	.0099	.0098	.0097	.0096	.0095	.0094	.0093	.0092
4.7	.0091	.0090	.0089	.0088	.0087	.0087	.0086	.0085	.0084	.0083
4.8	.0082	.0081	.0081	.0080	.0079	.0078	.0078	.0077	.0076	.0075
4.9	.0074	.0074	.0073	.0072	.0072	.0071	.0070	.0069	.0069	.0068
5.0	.0067									

Table 7 Percentage Points of the t Distribution

The table gives the value of $t_{\alpha,\nu}$ – the 100α percentage point of the t distribution for ν degrees of freedom.

The values of t are obtained by solution of the equation:

$$\alpha = \Gamma[\tfrac{1}{2}(\nu+1)][\Gamma(\tfrac{1}{2}\nu)]^{-1}(\nu\pi)^{-1/2}\int_{t}^{\infty}(1+x^2/\nu)^{-(\nu+1)/2}\,dx$$

Note: The tabulation is for one tail only, that is, for positive values of t. For $|t|$ the column headings for α should be doubled.

$\alpha =$	0.10	0.05	0.025	0.01	0.005	0.001	0.0005
$\nu = 1$	3.078	6.314	12.706	31.821	63.657	318.31	636.62
2	1.886	2.920	4.303	6.965	9.925	22.326	31.598
3	1.638	2.353	3.182	4.541	5.841	10.213	12.924
4	1.533	2.132	2.776	3.747	4.604	7.173	8.610
5	1.476	2.015	2.571	3.365	4.032	5.893	6.869
6	1.440	1.943	2.447	3.143	3.707	5.208	5.959
7	1.415	1.895	2.365	2.998	3.499	4.785	5.408
8	1.397	1.860	2.306	2.896	3.355	4.501	5.041
9	1.383	1.833	2.262	2.821	3.250	4.297	4.781
10	1.372	1.812	2.228	2.764	3.169	4.144	4.587
11	1.363	1.796	2.201	2.718	3.106	4.025	4.437
12	1.356	1.782	2.179	2.681	3.055	3.930	4.318
13	1.350	1.771	2.160	2.650	3.012	3.852	4.221
14	1.345	1.761	2.145	2.624	2.977	3.787	4.140
15	1.341	1.753	2.131	2.602	2.947	3.733	4.073
16	1.337	1.746	2.120	2.583	2.921	3.686	4.015
17	1.333	1.740	2.110	2.567	2.898	3.646	3.965
18	1.330	1.734	2.101	2.552	2.878	3.610	3.922
19	1.328	1.729	2.093	2.539	2.861	3.579	3.883
20	1.325	1.725	2.086	2.528	2.845	3.552	3.850
21	1.323	1.721	2.080	2.518	2.831	3.527	3.819
22	1.321	1.717	2.074	2.508	2.819	3.505	3.792
23	1.319	1.714	2.069	2.500	2.807	3.485	3.767
24	1.318	1.711	2.064	2.492	2.797	3.467	3.745
25	1.316	1.708	2.060	2.485	2.787	3.450	3.725
26	1.315	1.706	2.056	2.479	2.779	3.435	3.707
27	1.314	1.703	2.052	2.473	2.771	3.421	3.690
28	1.313	1.701	2.048	2.467	2.763	3.408	3.674
29	1.311	1.699	2.045	2.462	2.756	3.396	3.659
30	1.310	1.697	2.042	2.457	2.750	3.385	3.646
40	1.303	1.684	2.021	2.423	2.704	3.307	3.551
60	1.296	1.671	2.000	2.390	2.660	3.232	3.460
120	1.289	1.658	1.980	2.358	2.617	3.160	3.373
∞	1.282	1.645	1.960	2.326	2.576	3.090	3.291

This table is taken from Table III of Fisher & Yates: *Statistical Tables for Biological, Agricultural and Medical Research*, reprinted by permission of Addison Wesley Longman Ltd. Also from Table 12 of *Biometrika Tables for Statisticians*, Volume 1, by permission of Oxford University Press and the Biometrika Trustees.

Table 8 Percentage Points of the χ^2 Distribution

Table of $\chi^2_{\alpha,\nu}$ – the 100α percentage point of the χ^2 distribution for ν degrees of freedom.

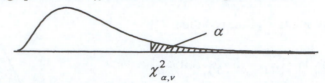

$$\chi^2_{\alpha,\nu}$$

$\alpha =$	.995	.99	.98	.975	.95	.90	.80	.75	.70	.50
$\nu = 1$	$.0^4393$	$.0^3157$	$.0^3628$	$.0^3982$	.00393	.0158	.0642	.102	.148	.455
2	.0100	.0201	.0404	.0506	.103	.211	.446	.575	.713	1.386
3	.0717	.115	.185	.216	.352	.584	1.005	1.213	1.424	2.366
4	.207	.297	.429	.484	.711	1.064	1.649	1.923	2.195	3.357
5	.412	.554	.752	.831	1.145	1.610	2.343	2.675	3.000	4.351
6	.676	.872	1.134	1.237	1.635	2.204	3.070	3.455	3.828	5.348
7	.989	1.239	1.564	1.690	2.167	2.833	3.822	4.255	4.671	6.346
8	1.344	1.646	2.032	2.180	2.733	3.490	4.594	5.071	5.527	7.344
9	1.735	2.088	2.532	2.700	3.325	4.168	5.380	5.899	6.393	8.343
10	2.156	2.558	3.059	3.247	3.940	4.865	6.179	6.737	7.267	9.342
11	2.603	3.053	3.609	3.816	4.575	5.578	6.989	7.584	8.148	10.341
12	3.074	3.571	4.178	4.404	5.226	6.304	7.807	8.438	9.034	11.340
13	3.565	4.107	4.765	5.009	5.892	7.042	8.634	9.299	9.926	12.340
14	4.075	4.660	5.368	5.629	6.571	7.790	9.467	10.165	10.821	13.339
15	4.601	5.229	5.985	6.262	7.261	8.547	10.307	11.036	11.721	14.339
16	5.142	5.812	6.614	6.908	7.962	9.312	11.152	11.912	12.624	15.338
17	5.697	6.408	7.255	7.564	8.672	10.085	12.002	12.792	13.531	16.338
18	6.265	7.015	7.906	8.231	9.390	10.865	12.857	13.675	14.440	17.338
19	6.844	7.633	8.567	8.907	10.117	11.651	13.716	14.562	15.352	18.338
20	7.434	8.260	9.237	9.591	10.851	12.443	14.578	15.452	16.266	19.337
21	8.034	8.897	9.915	10.283	11.591	13.240	15.445	16.344	17.182	20.337
22	8.643	9.542	10.600	10.982	12.338	14.041	16.314	17.240	18.101	21.337
23	9.260	10.196	11.293	11.688	13.091	14.848	17.187	18.137	19.021	22.337
24	9.886	10.856	11.992	12.401	13.848	15.659	18.062	19.037	19.943	23.337
25	10.520	11.524	12.697	13.120	14.611	16.473	18.940	19.939	20.867	24.337
26	11.160	12.198	13.409	13.844	15.379	17.292	19.820	20.843	21.792	25.336
27	11.808	12.879	14.125	14.573	16.151	18.114	20.703	21.749	22.719	26.336
28	12.461	13.565	14.847	15.308	16.928	18.939	21.588	22.657	23.647	27.336
29	13.121	14.256	15.574	16.047	17.708	19.768	22.475	23.567	24.577	28.336
30	13.787	14.953	16.306	16.791	18.493	20.599	23.364	24.478	25.508	29.336
40	20.706	22.164	23.838	24.433	26.509	29.051	32.345	33.660	34.872	39.335
50	27.991	29.707	31.664	32.357	34.764	37.689	41.449	42.942	44.313	49.335
60	35.535	37.485	39.699	40.482	43.188	46.459	50.641	52.294	53.809	59.335
70	43.275	45.442	47.893	48.758	51.739	55.329	59.898	61.698	63.346	69.334
80	51.171	53.539	56.213	57.153	60.391	64.278	69.207	71.145	72.915	79.334
90	59.196	61.754	64.634	65.646	69.126	73.291	78.558	80.625	82.511	89.334
100	67.327	70.065	73.142	74.222	77.929	82.358	87.945	90.133	92.129	99.334

For values of $\nu > 30$, approximate values of χ^2 may be obtained from the expression $\nu\left[1 - \dfrac{2}{9\nu} \pm \dfrac{x}{\sigma}\sqrt{\dfrac{2}{9\nu}}\right]^3$ where $\dfrac{x}{\sigma}$ is the normal deviate cutting off the corresponding tails of a normal distribution. If $\dfrac{x}{\sigma}$ is taken at the 0.02 level, so that 0.01 of the normal distribution is in each tail, the expression yields χ^2 at the 0.99 and 0.01 points.

For very large values of ν, it is sufficiently accurate to compute $\sqrt{2\chi^2}$, the distribution of which is approximately normal around a mean of $\sqrt{2\nu - 1}$ and with a standard deviation of 1.

Table taken from Table IV of Fisher and Yates: *Statistical Tables for Biological, Agricultural and Medical Research*, reprinted by permission of Addison Wesley Longman Ltd, and from Table 8 of *Biometrika Tables for Statisticians*, Vol. 1, by permission of Oxford University Press and the Biometrika Trustees.

Table 8 Percentage Points of the χ^2 Distribution – continued

.30	.25	.20	.10	.05	.025	.02	.01	.005	.001	$= \alpha$
1.074	1.323	1.642	2.706	3.841	5.024	5.412	6.635	7.879	10.827	$v = 1$
2.408	2.773	3.219	4.605	5.991	7.378	7.824	9.210	10.597	13.815	2
3.665	4.108	4.642	6.251	7.815	9.348	9.837	11.345	12.838	16.268	3
4.878	5.385	5.989	7.779	9.488	11.143	11.668	13.277	14.860	18.465	4
6.064	6.626	7.289	9.236	11.070	12.832	13.388	15.086	16.750	20.517	5
7.231	7.841	8.558	10.645	12.592	14.449	15.033	16.812	18.548	22.457	6
8.383	9.037	9.803	12.017	14.067	16.013	16.622	18.475	20.278	24.322	7
9.524	10.219	11.030	13.362	15.507	17.535	18.168	20.090	21.955	26.125	8
10.656	11.389	12.242	14.684	16.919	19.023	19.679	21.666	23.589	27.877	9
11.781	12.549	13.442	15.987	18.307	20.483	21.161	23.209	25.188	29.588	10
12.899	13.701	14.631	17.275	19.675	21.920	22.618	24.725	26.757	31.264	11
14.011	14.845	15.812	18.549	21.026	23.337	24.054	26.217	28.300	32.909	12
15.119	15.984	16.985	19.812	22.362	24.736	25.472	27.688	29.819	34.528	13
16.222	17.117	18.151	21.064	23.685	26.119	26.873	29.141	31.319	36.123	14
17.322	18.245	19.311	22.307	24.996	27.488	28.259	30.578	32.801	37.697	15
18.418	19.369	20.465	23.542	26.296	28.845	29.633	32.000	34.267	39.252	16
19.511	20.489	21.615	24.769	27.587	30.191	30.995	33.409	35.718	40.790	17
20.601	21.605	22.760	25.989	28.869	31.526	32.346	34.805	37.156	42.312	18
21.689	22.718	23.900	27.204	30.144	32.852	33.687	36.191	38.582	43.820	19
22.775	23.828	25.038	28.412	31.410	34.170	35.020	37.566	39.997	45.315	20
23.858	24.935	26.171	29.615	32.671	35.479	36.343	38.932	41.401	46.797	21
24.939	26.039	27.301	30.813	33.924	36.781	37.659	40.289	42.796	48.268	22
26.018	27.141	28.429	32.007	35.172	38.076	38.968	41.638	44.181	49.728	23
27.096	28.241	29.553	33.196	36.415	39.364	40.270	42.980	45.558	51.179	24
28.172	29.339	30.675	34.382	37.652	40.646	41.566	44.314	46.928	52.620	25
29.246	30.434	31.795	35.563	38.885	41.923	42.856	45.642	48.290	54.052	26
30.319	31.528	32.912	36.741	40.113	43.194	44.140	46.963	49.645	55.476	27
31.391	32.620	34.027	37.916	41.337	44.461	45.419	48.278	50.993	56.893	28
32.461	33.711	35.139	39.087	42.557	45.722	46.693	49.588	52.336	58.302	29
33.530	34.800	36.250	40.256	43.773	46.979	47.962	50.892	53.672	59.703	30
44.165	45.616	47.269	51.805	55.759	59.342	60.436	63.691	66.766	73.402	40
54.723	56.334	58.164	63.167	67.505	71.420	72.613	76.154	79.490	86.661	50
65.227	66.981	68.972	74.397	79.082	83.298	84.580	88.379	91.952	99.607	60
75.689	77.577	79.715	85.527	90.531	95.023	96.388	100.425	104.215	112.317	70
86.120	88.130	90.405	96.578	101.880	106.629	108.069	112.329	116.321	124.839	80
96.524	98.650	101.054	107.565	113.145	118.136	119.648	124.116	128.299	137.208	90
106.906	109.141	111.667	118.498	124.342	129.561	131.142	135.807	140.170	149.449	100

Table 9 Percentage Points of the F Distribution

The table gives the values of $F_{\alpha;\nu_1,\nu_2}$ the 100α percentage point of the F distribution having ν_1 degrees of freedom in the numerator and ν_2 degrees of freedom in the denominator. For each pair of values of ν_1 and ν_2, $F_{\alpha;\nu_1,\nu_2}$ is tabulated for $\alpha = 0.05$, 0.025, 0.01, 0.001, the 0.025 values being bracketed.

The lower percentage points of the distribution may be obtained from the relation:

$$F_{1-\alpha;\nu_1,\nu_2} = {}^{1}/F_{\alpha;\nu_2,\nu_1}$$

Example: $F_{.95,12,8} = {}^{1}/F_{.05,8,12} = {}^{1}/2.85 = \underline{0.351}$

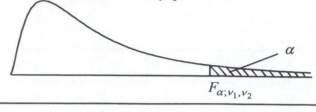

ν_1	1	2	3	4	5	6	7	8	10	12	24	∞
ν_2												
1	161.4	199.5	215.7	224.6	230.2	234.0	236.8	238.9	241.9	243.9	249.0	254.3
	(648)	(800)	(864)	(900)	(922)	(937)	(948)	(957)	(969)	(977)	(997)	(1018)
	4052	5000	5403	5625	5764	5859	5928	5981	6056	6106	6235	6366
	4053*	5000*	5405*	5625*	5764*	5859*	5929*	5981*	6056*	6107*	6235*	6366*
2	18.5	19.0	19.2	19.2	19.3	19.3	19.4	19.4	19.4	19.4	19.5	19.5
	(38.5)	(39.0)	(39.2)	(39.2)	(39.3)	(39.3)	(39.4)	(39.4)	(39.4)	(39.4)	(39.5)	(39.5)
	98.5	99.0	99.2	99.2	99.3	99.3	99.4	99.4	99.4	99.4	99.5	99.5
	998.5	999.0	999.2	999.2	999.3	999.3	999.4	999.4	999.4	999.4	999.5	999.5
3	10.13	9.55	9.28	9.12	9.01	8.94	8.89	8.85	8.79	8.74	8.64	8.53
	(17.4)	(16.0)	(15.4)	(15.1)	(14.9)	(14.7)	(14.6)	(14.5)	(14.4)	(14.3)	(14.1)	(13.9)
	34.1	30.8	29.5	28.7	28.2	27.9	27.7	27.5	27.2	27.1	26.6	26.1
	167.0	148.5	141.1	137.1	134.6	132.8	131.5	130.6	129.2	128.3	125.9	123.5
4	7.71	6.94	6.59	6.39	6.26	6.16	6.09	6.04	5.96	5.91	5.77	5.63
	(12.22)	(10.65)	(9.98)	(9.60)	(9.36)	(9.20)	(9.07)	(8.98)	(8.84)	(8.75)	(8.51)	(8.26)
	21.2	18.0	16.7	16.0	15.5	15.2	15.0	14.8	14.5	14.4	13.9	13.5
	74.14	61.25	56.18	53.44	51.71	50.53	49.66	49.00	48.05	47.41	45.77	44.05
5	6.61	5.79	5.41	5.19	5.05	4.95	4.88	4.82	4.74	4.68	4.53	4.36
	(10.01)	(8.43)	(7.76)	(7.39)	(7.15)	(6.98)	(6.85)	(6.76)	(6.62)	(6.52)	(6.28)	(6.02)
	16.26	13.27	12.06	11.39	10.97	10.67	10.46	10.29	10.05	9.89	9.47	9.02
	47.18	37.12	33.20	31.09	29.75	28.83	28.16	27.65	26.92	26.42	25.14	23.79
6	5.99	5.14	4.76	4.53	4.39	4.28	4.21	4.15	4.06	4.00	3.84	3.67
	(8.81)	(7.26)	(6.60)	(6.23)	(5.99)	(5.82)	(5.70)	(5.60)	(5.46)	(5.37)	(5.12)	(4.85)
	13.74	10.92	9.78	9.15	8.75	8.47	8.26	8.10	7.87	7.72	7.31	6.88
	35.51	27.00	23.70	21.92	20.80	20.03	19.46	19.03	18.41	17.99	16.90	15.75
7	5.59	4.74	4.35	4.12	3.97	3.87	3.79	3.73	3.64	3.57	3.41	3.23
	(8.07)	(6.54)	(5.89)	(5.52)	(5.29)	(5.12)	(4.99)	(4.90)	(4.76)	(4.67)	(4.42)	(4.14)
	12.25	9.55	8.45	7.85	7.46	7.19	6.99	6.84	6.62	6.47	6.07	5.65
	29.25	21.69	18.77	17.20	16.21	15.52	15.02	14.63	14.08	13.71	12.73	11.70
8	5.32	4.46	4.07	3.84	3.69	3.58	3.50	3.44	3.35	3.28	3.12	2.93
	(7.57)	(6.06)	(5.42)	(5.05)	(4.82)	(4.65)	(4.53)	(4.43)	(4.30)	(4.20)	(3.95)	(3.67)
	11.26	8.65	7.59	7.01	6.63	6.37	6.18	6.03	5.81	5.67	5.28	4.86
	25.42	18.49	15.83	14.39	13.48	12.86	12.40	12.05	11.54	11.19	10.30	9.34
9	5.12	4.26	3.86	3.63	3.48	3.37	3.29	3.23	3.14	3.07	2.90	2.71
	(7.21)	(5.71)	(5.08)	(4.72)	(4.48)	(4.32)	(4.20)	(4.10)	(3.96)	(3.87)	(3.61)	(3.33)
	10.56	8.02	6.99	6.42	6.06	5.80	5.61	5.47	5.26	5.11	4.73	4.31
	22.86	16.39	13.90	12.56	11.71	11.13	10.69	10.37	9.87	9.57	8.72	7.81
10	4.96	4.10	3.71	3.48	3.33	3.22	3.14	3.07	2.98	2.91	2.74	2.54
	(6.94)	(5.46)	(4.83)	(4.47)	(4.24)	(4.07)	(3.95)	(3.85)	(3.72)	(3.62)	(3.37)	(3.08)
	10.04	7.56	6.55	5.99	5.64	5.39	5.20	5.06	4.85	4.71	4.33	3.91
	21.04	14.91	12.55	11.28	10.48	9.93	9.52	9.20	8.74	8.44	7.64	6.76
11	4.84	3.98	3.59	3.36	3.20	3.09	3.01	2.95	2.85	2.79	2.61	2.40
	(6.72)	(5.26)	(4.63)	(4.28)	(4.04)	(3.88)	(3.76)	(3.66)	(3.53)	(3.43)	(3.17)	(2.88)
	9.65	7.21	6.22	5.67	5.32	5.07	4.89	4.74	4.54	4.40	4.02	3.60
	19.69	13.81	11.56	10.35	9.58	9.05	8.66	8.35	7.92	7.63	6.85	6.00
12	4.75	3.89	3.49	3.26	3.11	3.00	2.91	2.85	2.75	2.69	2.51	2.30
	(6.55)	(5.10)	(4.47)	(4.12)	(3.89)	(3.73)	(3.61)	(3.51)	(3.37)	(3.28)	(3.02)	(2.72)
	9.33	6.93	5.95	5.41	5.06	4.82	4.64	4.50	4.30	4.16	3.78	3.36
	18.64	12.97	10.80	9.63	8.89	8.38	8.00	7.71	7.29	7.00	6.25	5.42

* Entries marked thus must be multiplied by 100

Table 9 Percentage Points of the *F* Distribution - continued

ν_1	1	2	3	4	5	6	7	8	10	12	24	∞
ν_2												
13	4.67	3.81	3.41	3.18	3.03	2.92	2.83	2.77	2.67	2.60	2.42	2.21
	(6.41)	(4.97)	(4.35)	(4.00)	(3.77)	(3.60)	(3.48)	(3.39)	(3.25)	(3.15)	(2.89)	(2.60)
	9.07	6.70	5.74	5.21	4.86	4.62	4.44	4.30	4.10	3.96	3.59	3.17
	17.82	12.31	10.21	9.07	8.35	7.86	7.49	7.21	6.80	6.52	5.78	4.97
14	4.60	3.74	3.34	3.11	2.96	2.85	2.76	2.70	2.60	2.53	2.35	2.13
	(6.30)	(4.86)	(4.24)	(3.89)	(3.66)	(3.50)	(3.38)	(3.29)	(3.15)	(3.05)	(2.79)	(2.49)
	8.86	6.51	5.56	5.04	4.70	4.46	4.28	4.14	3.94	3.80	3.43	3.00
	17.14	11.78	9.73	8.62	7.92	7.44	7.08	6.80	6.40	6.13	5.41	4.60
16	4.49	3.63	3.24	3.01	2.85	2.74	2.66	2.59	2.49	2.42	2.24	2.01
	(6.12)	(4.69)	(4.08)	(3.73)	(3.50)	(3.34)	(3.22)	(3.12)	(2.99)	(2.89)	(2.63)	(2.32)
	8.53	6.23	5.29	4.77	4.44	4.20	4.03	3.89	3.69	3.55	3.18	2.75
	16.12	10.97	9.01	7.94	7.27	6.80	6.46	6.19	5.81	5.55	4.85	4.06
18	4.41	3.55	3.16	2.93	2.77	2.66	2.58	2.51	2.41	2.34	2.15	1.92
	(5.98)	(4.56)	(3.95)	(3.61)	(3.38)	(3.22)	(3.10)	(3.01)	(2.87)	(2.77)	(2.50)	(2.19)
	8.29	6.01	5.09	4.58	4.25	4.01	3.84	3.71	3.51	3.37	3.00	2.57
	15.38	10.39	8.49	7.46	6.81	6.35	6.02	5.76	5.39	5.13	4.45	3.67
20	4.35	3.49	3.10	2.87	2.71	2.60	2.51	2.45	2.35	2.28	2.08	1.84
	(5.87)	(4.46)	(3.86)	(3.51)	(3.29)	(3.13)	(3.01)	(2.91)	(2.77)	(2.68)	(2.41)	(2.09)
	8.10	5.85	4.94	4.43	4.10	3.87	3.70	3.56	3.37	3.23	2.86	2.42
	14.82	9.95	8.10	7.10	6.46	6.02	5.69	5.44	5.08	4.82	4.15	3.38
22	4.30	3.44	3.05	2.82	2.66	2.55	2.46	2.40	2.30	2.23	2.03	1.78
	(5.79)	(4.38)	(3.78)	(3.44)	(3.22)	(3.05)	(2.93)	(2.84)	(2.70)	(2.60)	(2.33)	(2.00)
	7.95	5.72	4.82	4.31	3.99	3.76	3.59	3.45	3.26	3.12	2.75	2.31
	14.38	9.61	7.80	6.81	6.19	5.76	5.44	5.19	4.83	4.58	3.92	3.15
24	4.26	3.40	3.01	2.78	2.62	2.51	2.42	2.36	2.25	2.18	1.98	1.73
	(5.72)	(4.32)	(3.72)	(3.38)	(3.15)	(2.99)	(2.87)	(2.78)	(2.64)	(2.54)	(2.27)	(1.94)
	7.82	5.61	4.72	4.22	3.90	3.67	3.50	3.36	3.17	3.03	2.66	2.21
	14.03	9.34	7.55	6.59	5.98	5.55	5.23	4.99	4.64	4.39	3.74	2.97
26	4.23	3.37	2.98	2.74	2.59	2.47	2.39	2.32	2.22	2.15	1.95	1.69
	(5.66)	(4.27)	(3.67)	(3.33)	(3.10)	(2.94)	(2.82)	(2.73)	(2.59)	(2.49)	(2.22)	(1.88)
	7.72	5.53	4.64	4.14	3.82	3.59	3.42	3.29	3.09	2.96	2.58	2.13
	13.74	9.12	7.36	6.41	5.80	5.38	5.07	4.83	4.48	4.24	3.59	2.82
28	4.20	3.34	2.95	2.71	2.56	2.45	2.36	2.29	2.19	2.12	1.91	1.65
	(5.61)	(4.22)	(3.63)	(3.29)	(3.06)	(2.90)	(2.78)	(2.69)	(2.55)	(2.45)	(2.17)	(1.83)
	7.64	5.45	4.57	4.07	3.75	3.53	3.36	3.23	3.03	2.90	2.52	2.06
	13.50	8.93	7.19	6.25	5.66	5.24	4.93	4.69	4.35	4.11	3.46	2.69
30	4.17	3.32	2.92	2.69	2.53	2.42	2.33	2.27	2.16	2.09	1.89	1.62
	(5.57)	(4.18)	(3.59)	(3.25)	(3.03)	(2.87)	(2.75)	(2.65)	(2.51)	(2.41)	(2.14)	(1.79)
	7.56	5.39	4.51	4.02	3.70	3.47	3.30	3.17	2.98	2.84	2.47	2.01
	13.29	8.77	7.05	6.12	5.53	5.12	4.82	4.58	4.24	4.00	3.36	2.59
40	4.08	3.23	2.84	2.61	2.45	2.34	2.25	2.18	2.08	2.00	1.79	1.51
	(5.42)	(4.05)	(3.46)	(3.13)	(2.90)	(2.74)	(2.62)	(2.53)	(2.39)	(2.29)	(2.01)	(1.64)
	7.31	5.18	4.31	3.83	3.51	3.29	3.12	2.99	2.80	2.66	2.29	1.80
	12.61	8.25	6.59	5.70	5.13	4.73	4.44	4.21	3.87	3.64	3.01	2.23
60	4.00	3.15	2.76	2.53	2.37	2.25	2.17	2.10	1.99	1.92	1.70	1.39
	(5.29)	(3.93)	(3.34)	(3.01)	(2.79)	(2.63)	(2.51)	(2.41)	(2.27)	(2.17)	(1.88)	(1.48)
	7.08	4.98	4.13	3.65	3.34	3.12	2.95	2.82	2.63	2.50	2.12	1.60
	11.97	7.77	6.17	5.31	4.76	4.37	4.09	3.86	3.54	3.32	2.69	1.89
120	3.92	3.07	2.68	2.45	2.29	2.18	2.09	2.02	1.91	1.83	1.61	1.25
	(5.15)	(3.80)	(3.23)	(2.89)	(2.67)	(2.52)	(2.39)	(2.30)	(2.16)	(2.05)	(1.76)	(1.31)
	6.85	4.79	3.95	3.48	3.17	2.96	2.79	2.66	2.47	2.34	1.95	1.38
	11.38	7.32	5.78	4.95	4.42	4.04	3.77	3.55	3.24	3.02	2.40	1.54
∞	3.84	3.00	2.60	2.37	2.21	2.10	2.01	1.94	1.83	1.75	1.52	1.00
	(5.02)	(3.69)	(3.12)	(2.79)	(2.57)	(2.41)	(2.29)	(2.19)	(2.05)	(1.94)	(1.64)	(1.00)
	6.63	4.61	3.78	3.32	3.02	2.80	2.64	2.51	2.32	2.18	1.79	1.00
	10.83	6.91	5.42	4.62	4.10	3.74	3.47	3.27	2.96	2.74	2.13	1.00

This table is taken from Table V of Fisher & Yates: *Statistical Tables for Biological, Agricultural and Medical Research*, reprinted by permission of Addison Wesley Longman Ltd and also from Table 18 of *Biometrika Tables for Statisticians*, Volume 1, by permission of Oxford University Press and the Biometrika Trustees.

Table 10 Percentage Points of the Correlation Coefficient

Values of the Correlation Coefficient for Different Levels of Significance

For a total correlation, v is 2 less than the number of pairs in the sample; for a partial correlation, the number of eliminated variates should also be subtracted. The probabilities at the head of the columns refer to the two-tail test of significance and give the chance that $|r|$ will be greater than the tabulated values given that the true product moment correlation, ρ, is zero. For a single-tail test the probabilities should be halved.

Example: In a test for a significant positive correlation between two variables, the observed correlation coefficient of ten pairs of observations would have to exceed the value of +0.7155 to be significant at the 1% level.

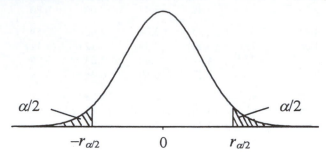

	0.1	0.05	0.02	0.01	0.001
$v = 1$	.98769	.99692	.999507	.999877	.9999988
2	.90000	.95000	.98000	.990000	.99900
3	.8054	.8783	.93433	.95873	.99116
4	.7293	.8114	.8822	.91720	.97406
5	.6694	.7545	.8329	.8745	.95074
6	.6215	.7067	.7887	.8343	.92493
7	.5822	.6664	.7498	.7977	.8982
8	.5494	.6319	.7155	.7646	.8721
9	.5214	.6021	.6851	.7348	.8471
10	.4973	.5760	.6581	.7079	.8233
11	.4762	.5529	.6339	.6835	.8010
12	.4575	.5324	.6120	.6614	.7800
13	.4409	.5139	.5923	.6411	.7603
14	.4259	.4973	.5742	.6226	.7420
15	.4124	.4821	.5577	.6055	.7246
16	.4000	.4683	.5425	.5897	.7084
17	.3887	.4555	.5285	.5751	.6932
18	.3783	.4438	.5155	.5614	.6787
19	.3687	.4329	.5034	.5487	.6652
20	.3598	.4227	.4921	.5368	.6524
25	.3233	.3809	.4451	.4869	.5974
30	.2960	.3494	.4093	.4487	.5541
35	.2746	.3246	.3810	.4182	.5189
40	.2573	.3044	.3578	.3932	.4896
45	.2428	.2875	.3384	.3721	.4648
50	.2306	.2732	.3218	.3541	.4433
60	.2108	.2500	.2948	.3248	.4078
70	.1954	.2319	.2737	.3017	.3799
80	.1829	.2172	.2565	.2830	.3568
90	.1726	.2050	.2422	.2673	.3375
100	.1638	.1946	.2301	.2540	.3211

This is table is taken from Table VII of Fisher and Yates: *Statistical Tables for Biological, Agricultural and Medical Research*, reprinted by permission of Addison Wesley Longman Ltd.

Table 11 Tukey's Wholly Significant Difference (5% Level)

This table gives the values of the WSD/$s(\bar{x})$ where:

WSD	= 'Wholly Significant Difference' between means at 5% level
s	= Estimate of standard deviation
$s(\bar{x})$	= Estimate of standard error of the means = $s/\sqrt{n}$
k	= Number of means under test
v	= Degrees of freedom associated with s^2, the estimate of error variance
n	= Number of readings in each mean

k v	2	3	4	5	6	7	8	9	10	12	15	20	30	60
1	21.96	28.80	34.56	39.60	44.28	48.78	53.10	55.62	57.96	62.46	69.12	77.58	91.98	116.64
2	6.83	8.54	9.88	10.98	11.95	12.87	13.73	14.27	14.76	15.68	17.02	18.67	21.41	26.11
3	4.89	6.00	6.84	7.50	8.07	8.60	9.08	9.41	9.69	10.22	10.98	11.89	13.40	15.96
4	4.19	5.10	5.76	6.27	6.70	7.10	7.45	7.70	7.92	8.31	8.88	9.55	10.62	12.47
5	3.86	4.66	5.24	5.68	6.04	6.37	6.66	6.88	7.06	7.39	7.86	8.41	9.28	10.77
6	3.64	4.38	4.91	5.31	5.63	5.92	6.17	6.37	6.53	6.82	7.23	7.70	8.45	9.73
7	3.50	4.20	4.70	5.06	5.35	5.62	5.85	6.03	6.17	6.44	6.81	7.24	7.91	9.05
8	3.41	4.08	4.55	4.89	5.17	5.41	5.62	5.79	6.00	6.18	6.53	6.92	7.53	8.57
9	3.34	3.98	4.44	4.76	5.03	5.26	5.46	5.62	5.75	5.98	6.31	6.68	7.25	7.99
10	3.28	3.91	4.35	4.66	4.91	5.13	5.32	5.48	5.61	5.83	6.14	6.49	7.02	7.94
12	3.19	3.80	4.22	4.52	4.75	4.96	5.13	5.28	5.40	5.61	5.90	6.22	6.70	7.54
15	3.11	3.69	4.09	4.37	4.60	4.79	4.95	5.09	5.20	5.39	5.66	5.95	6.39	7.14
20	3.04	3.60	3.98	4.25	4.45	4.63	4.78	4.91	5.02	5.19	5.44	5.71	6.11	6.79
30	2.97	3.51	3.87	4.12	4.32	4.48	4.61	4.74	4.84	5.00	5.23	5.47	5.83	6.44
60	2.90	3.41	3.76	4.00	4.18	4.33	4.45	4.57	4.66	4.81	5.02	5.24	5.56	6.09
∞	2.82	3.32	3.66	3.88	4.04	4.18	4.29	4.40	4.49	4.63	4.82	5.01	5.29	5.76

See page 72 for an example of the use of this table.

Table from W. Volk, *Applied Statistics for Engineers*, McGraw-Hill, 1969, p 179, (Modification of tables in *The Problem of Multiple Comparison*, mimeographed publication 1953, J. W. Tukey), reproduced with permission of The McGraw-Hill Companies.

Table 12 Percentage Points of Spearman's Rank Correlation Coefficient

The coefficient r_s is calculated as $1 - \dfrac{6\Sigma d^2}{n(n^2-1)}$ where n is the number of observations in each of two series and d is the difference between the ranks of the corresponding observations in each series.

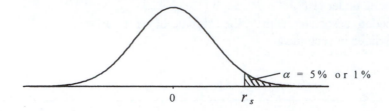

An observed value of r_s greater than or equal to that tabulated indicates a significant positive correlation between the ranks of the two series of observations at the level indicated (5% or 1%). For a two-sided test the probabilities should be doubled. For a significant negative correlation between ranks, an observed r_s should be negative but numerically greater than or equal to the tabulated value.

	(One-tailed test)	
$\alpha =$	5%	1%
$n = 4$	1.000	–
5	0.900	1.000
6	0.829	0.943
7	0.714	0.893
8	0.643	0.833
9	0.600	0.783
10	0.564	0.746
12	0.506	0.712
14	0.456	0.645
16	0.425	0.601
18	0.399	0.564
20	0.377	0.534
22	0.359	0.508
24	0.343	0.485
26	0.329	0.465
28	0.317	0.448
30	0.306	0.432

When n is large (greater than about 10 say), the significance of an observed value of r_s under the null hypothesis may be tested using:

$$t \approx r_s \sqrt{\frac{n-2}{1-r_s^2}}$$

where t is Student's t with $n-2$ degrees of freedom.

See page 73 for an example of the use of this table.

Adapted from Olds E.G. 1938 Ann. Math. Statist. 9, 133-148 and Olds E.G. 1949 Ann. Math. Statist., 20, 117-118. Reprinted with permission of the Institute of Mathematical Statistics.

Table 13 Percentage Points of Kendall's Rank Correlation Coefficient

The coefficient r_K is calculated as $\dfrac{S}{\dfrac{n}{2}(n-1)}$ where n is the number of observations in each of two series.

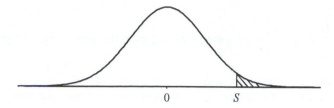

Note that the tabulation here is in terms of the numerator S which, for a given n, has a one-to-one correspondence with r_K.

The paired observations of ranks are rearranged so that one set is in ascending order. The corresponding ranks of the second set are examined two at a time, scoring $+1$ if the pair is in ascending order and -1 if it is in descending order. S is the net score for the $\dfrac{n}{2}(n-1)$ pairs amongst the n ranks of the second set.

The table gives the probability that the net total score, S, will be greater than or equal to any possible positive values when the ranks of the two sets are not correlated. Since the distribution of S, and also of r_K, is symmetrical, a negative correlation of ranks can be tested against the null hypothesis by interpreting the tabulated S values as negative. For a two-tailed test, the tabulated probability corresponding to the observed value of S should be doubled (except for $S = 0$).

S	4	5	8	9		S	6	7	10
0	.625	.592	.548	.540		1	.500	.500	.500
2	.375	.408	.452	.460		3	.360	.386	.431
4	.167	.242	.360	.381		5	.235	.281	.364
6	.042	.117	.274	.306		7	.136	.191	.300
8		.042	.199	.238		9	.068	.119	.242
10		.0083	.138	.179		11	.028	.068	.190
12			.089	.130		13	.0083	.035	.146
14			.054	.090		15	.0014	.015	.108
16			.031	.060		17		.0054	.078
18			.016	.038		19		.0014	.054
20			.0071	.022		21		.0002	.036
22			.0028	.012		23			.023
24			.0009	.0063		25			.014
26			.0002	.0029		27			.0083
28				.0012		29			.0046
30				.0004		31			.0023
						33			.0011
						35			.0005

For values of n greater than 10, it is sufficiently accurate to use the fact that r_K is approximately normal with mean of zero and variance of:

$$\frac{2(2n+5)}{9n(n-1)}$$

Thus r_K can be expressed as a standardised normal variate by writing:

$$z = \frac{\dfrac{S}{\dfrac{n}{2}(n-1)} - 0}{\sqrt{\left(\dfrac{2(2n+5)}{9n(n-1)}\right)}} = S\sqrt{\frac{18}{n(n-1)(2n+5)}}$$

The significance of S can then be tested using tables of the normal distribution. See page 74 for an example of the use of this table.

Adapted from Kendall M.G., *Rank Correlation Methods*, Appendix Table 1, Charles Griffin, 1948, reprinted with permission of Edward Arnold (Publishers) Ltd.

Table 14 Percentage Points of Nair's 'Studentised' Extreme Deviate from the Mean

For a random sample of size n from a normal population, the lowest observation is $x_{(1)}$, the highest is $x_{(n)}$ and the arithmetic mean is $\bar{x}$. s is an independent estimate, external to the sample, of the population standard deviation and having ν degrees of freedom.

The maximum value of $\dfrac{|x_{(1)} - \bar{x}|}{s}$ and $\dfrac{|x_{(n)} - \bar{x}|}{s}$ gives a test criterion for identifying a single possible extreme value.

The table gives the upper 5% and 1% points of the distribution of the statistic for n and ν. An observed value of the statistic greater than or equal to that tabulated suggests that $x_{(1)}$ or $x_{(n)}$ as appropriate is not from the same population as the remainder of the sample.

			5% Significance Level				
$n =$	3	4	5	6	7	8	9
$\nu = 10$	2.02	2.29	2.49	2.63	2.75	2.85	2.93
11	1.99	2.26	2.44	2.58	2.70	2.79	2.87
12	1.97	2.22	2.40	2.54	2.65	2.75	2.83
13	1.95	2.20	2.38	2.51	2.62	2.71	2.79
14	1.93	2.18	2.35	2.48	2.59	2.68	2.76
15	1.92	2.16	2.33	246	2.56	2.65	2.73
16	1.90	2.14	2.31	2.44	2.54	2.63	2.70
17	1.89	2.13	2.30	2.42	2.52	2.61	2.68
18	1.88	2.12	2.28	2.41	2.51	2.59	2.66
19	1.87	2.11	2.27	2.39	2.49	2.58	2.65
20	1.87	2.10	2.26	2.38	2.48	2.56	2.63
24	1.84	2.07	2.23	2.35	2.44	2.52	2.59
30	1.82	2.04	2.20	2.31	2.40	2.48	2.55
40	1.80	2.02	2.17	2.28	2.37	2.44	2.51
60	1.78	1.99	2.14	2.25	2.33	2.41	2.47
120	1.76	1.97	2.11	2.21	2.30	2.37	2.43
∞	1.74	1.94	2.08	2.18	2.27	2.33	2.39

			1% Significance Level				
$n =$	3	4	5	6	7	8	9
$\nu = 10$	2.76	3.05	3.25	3.39	3.50	3.59	3.67
11	2.71	3.00	3.19	3.33	3.44	3.53	3.61
12	2.67	2.95	3.14	3.28	3.39	3.48	3.55
13	2.63	2.91	3.10	3.24	3.34	3.43	3.51
14	2.60	2.87	3.06	3.20	3.30	3.39	3.47
15	2.57	2.84	3.02	3.16	3.27	3.35	3.43
16	2.55	2.81	3.00	3.13	3.24	3.32	3.39
17	2.53	2.79	2.97	3.10	3.21	3.29	3.36
18	2.50	2.77	2.95	3.08	3.18	3.27	3.34
19	2.49	2.75	2.92	3.06	3.16	3.24	3.31
20	2.47	2.73	2.91	3.04	3.14	3.22	3.29
24	2.43	2.68	2.85	2.97	3.07	3.15	3.22
30	2.38	2.62	2.79	2.91	3.01	3.08	3.15
40	2.34	2.57	2.73	2.85	2.94	3.02	3.08
60	2.30	2.52	2.68	2.79	2.88	2.95	3.01
120	2.25	2.48	2.62	2.73	2.82	2.89	2.95
∞	2.22	2.43	2.57	2.68	2.76	2.83	2.88

See page 75 for an example on the use of this table.

Table taken from K.R. Nair, *Biometrika*, Vol. 39, 1952, reprinted by permission of Oxford University Press and the Biometrika Trustees.

Table 15 Upper Percentage Points of Dixon's Rank Difference Ratio

The table gives the upper 100α percentage points of three rank difference ratios, each ratio being appropriate to a particular range of sample size, $\alpha = 0.10$, 0.05 and 0.01. No estimate of the population standard deviation is necessary.

If a random sample of size n from a normal population is ranked in ascending order of magnitude resulting in $x_{(1)}$, $x_{(2)}$, ... $x_{(n)}$, then an observed value of the rank difference ratio greater than or equal to that tabulated indicates that the particular extreme value of x, that is $x_{(1)}$ or $x_{(n)}$, is not consistent with the remainder of the sample.

Rank Difference Ratio			Significance Level		
		$\alpha =$	.010	0.05	0.01
$\dfrac{x_{(2)} - x_{(1)}}{x_{(n)} - x_{(1)}}$ or $\dfrac{x_{(n)} - x_{(n-1)}}{x_{(n)} - x_{(1)}}$		$n = 3$	0.886	0.941	0.988
		4	0.679	0.765	0.889
		5	0.557	0.642	0.780
		6	0.482	0.560	0.698
		7	0.434	0.507	0.637
$\dfrac{x_{(3)} - x_{(1)}}{x_{(n-1)} - x_{(1)}}$ or $\dfrac{x_{(n)} - x_{(n-2)}}{x_{(n)} - x_{(2)}}$		8	0.650	0.710	0.829
		9	0.594	0.657	0.776
		10	0.551	0.612	0.726
		11	0.517	0.576	0.679
		12	0.490	0.546	0.642
		13	0.467	0.521	0.615
		14	0.448	0.501	0.593
$\dfrac{x_{(3)} - x_{(1)}}{x_{(n-2)} - x_{(1)}}$ or $\dfrac{x_{(n)} - x_{(n-2)}}{x_{(n)} - x_{(3)}}$		15	0.472	0.525	0.616
		16	0.454	0.507	0.595
		17	0.438	0.490	0.577
		18	0.424	0.475	0.561
		19	0.412	0.462	0.547
		20	0.401	0.450	0.535

See page 75 for an example on the use of this table.

Data taken from W.J. Dixon, *Ann. Math. Statis.*, 22: 68 (1951). Reprinted with permission of the Institute of Mathematical Statistics.

Table 16 Percentage Points of D in the One-sample Kolmogorov-Smirnov Distribution

The table gives the values of D_α, the 100α percentage point of the D distribution for α = 0.20, 0.15, 0.10, 0.05 and 0.01.

D is the maximum value of $|F_{OBS}(x) - F_{EXP}(x)|$ where $F_{OBS}(x)$ and $F_{EXP}(x)$ are observed and expected (theoretical) cumulative probability distribution functions evaluated at each of a set of values of the observed variable x.

Values of D greater than or equal to those tabulated for the given sample size n and significance level α suggest that the sample has not been drawn from a population with the properties of the assumed theoretical distribution.

n \\ α	.20	.15	.10	.05	.01
1	.900	.925	.950	.975	.995
2	.684	.726	.776	.842	.929
3	.565	.597	.642	.708	.828
4	.494	.525	.564	.624	.733
5	.446	.474	.510	.565	.669
6	.410	.436	.470	.521	.618
7	.381	.405	.438	.486	.577
8	.358	.381	.411	.457	.543
9	.339	.360	.388	.432	.514
10	.322	.342	.368	.410	.490
11	.307	.326	.352	.391	.468
12	.295	.313	.338	.375	.450
13	.284	.302	.325	.361	.433
14	.274	.292	.314	.349	.418
15	.266	.283	.304	.338	.404
16	.258	.274	.295	.328	.392
17	.250	.266	.286	.318	.381
18	.244	.259	.278	.309	.371
19	.237	.252	.272	.301	.363
20	.231	.246	.264	.294	.356
25	.21	.22	.24	.27	.32
30	.19	.20	.22	.24	.29
35	.18	.19	.21	.23	.27
Over 35	$\dfrac{1.07}{\sqrt{n}}$	$\dfrac{1.14}{\sqrt{n}}$	$\dfrac{1.22}{\sqrt{n}}$	$\dfrac{1.36}{\sqrt{n}}$	$\dfrac{1.63}{\sqrt{n}}$

See page 76 for an example of the use of this table.

Adapted from Massey, F. J., Jr. 1951. 'The Kolmogorov-Smirnov test for goodness of fit'. *Journal of the American Statistical Association*, vol. 46, 68-78.

Table 17 Lower Percentage Points of the Wilcoxon Signed-rank Distribution

For a random sample from a symmetric continuous distribution with mean of μ_0, subtract μ_0 from each sample member, discard all zero values and rank the *absolute* values of the remaining differences in ascending order. For tied ranks, use the average of the ranks if there had been no ties. Let w_+ and w_- be the sum of the ranks which correspond to positive and negative differences respectively.

The table gives the lower percentage points of W_+, the sum of positive ranks. A test of the hypothesis that a sample has come from a continuous symmetric population with specified mean value μ_0 (or that the mean difference of matched pairs is μ_0) against the alternative hypothesis that $\mu < \mu_0$, would lead to rejection if the calculated value of w_+ is less than or equal to the tabulated value of W_+. To test the hypothesis against the alternative $\mu > \mu_0$, reject if w_- is less than or equal to the tabulated value. For the two-sided alternative $\mu \neq \mu_0$, use the smaller value of w_+ and w_-; in this case the tabulated value of α should be doubled.

Because the distribution of ranks is discrete, the possible values of W are not usually associated with round number values of probability (as usually quoted for α). The table gives the exact probability of observing a given value of W_+ or smaller. This exact probability is the one *nearer* to the given α, not the largest one which is less than or equal to α. Where this probability is unacceptably larger than required, reduction of the tabulated value of W_+ by 1 will bring the significance level below the designated value of α. For $\alpha = 0.001$, the probabilities of the given W_+ are all less than or equal to 0.001.

α	0.05	Nearer exact probability	0.025	Nearer exact probability	0.01	Nearer exact probability	0.005	Nearer exact probability	0.001
$n = 5$	1	.0625	–		–		–		–
6	2	.0469	1	.0312	–		–		–
7	4	.0547	2	.0234	0	.0078	–		–
8	6	.0547	4	.0273	2	.0117	0	.0039	–
9	8	.0488	6	.0273	3	.0098	2	.0059	–
10	11	.0527	8	.0244	5	.0098	3	.0049	0
11	14	.0508	11	.0269	7	.0093	5	.0049	1
12	17	.0461	14	.0261	10	.0105	7	.0046	2
13	21	.0471	17	.0239	13	.0107	10	.0052	4
14	26	.0520	21	.0247	16	.0101	13	.0054	6
15	30	.0473	25	.0240	20	.0108	16	.0051	8
16	36	.0523	30	.0253	24	.0107	20	.0055	11
17	41	.0492	35	.0253	28	.0101	23	.0047	14
18	47	.0494	40	.0241	33	.0104	28	.0052	18
19	54	.0521	46	.0247	38	.0102	32	.0047	21
20	60	.0487	52	.0242	43	.0096	38	.0053	26
21	68	.0516	59	.0251	49	.0097	43	.0051	30
22	75	.0492	66	.0250	56	.0104	49	.0052	35
23	83	.0490	73	.0242	62	.0098	55	.0051	40
24	92	.0505	81	.0245	69	.0097	61	.0048	45
25	101	.0507	90	.0258	77	.0101	68	.0048	51
26	110	.0497	98	.0247	85	.0102	76	.0051	58
27	120	.0502	107	.0246	93	.0100	84	.0052	64
28	130	.0496	117	.0252	102	.0102	92	.0051	71
29	141	.0504	127	.0253	111	.0101	100	.0049	79
30	152	.0502	137	.0249	120	.0098	109	.0050	86

As n increases, the distribution of W_+ (and W_-) tends to normality with mean of $\dfrac{n(n+1)}{4}$ and standard deviation of $\sqrt{\dfrac{n(n+1)(2n+1)}{24}}$

For $n > 30$, the use of the approximation will give results accurate enough for most practical purposes.

Table 18 Percentage Points of D in the Two-sample Kolmogorov-Smirnov Distribution

The table gives the values of D_α, the 100α upper percentage points of the Kolmogorov-Smirnov two-sample test statistic for $\alpha = 0.05, 0.025, 0.01$ and 0.001.

In the two-sample case, given two random samples of size n_1 and n_2 respectively from the same distribution of a continuous variate, x, D is defined as the maximum absolute value of $F_1(x) - F_2(x)$.

$F_1(x)$ and $F_2(x)$ are the two empirical cumulative relative frequency distribution functions evaluated at each of a set of values of the observed variable, x. It is recommended that the number of values of x, and hence the number of values of $|F_1(x) - F_2(x)|$, should be as large as feasible given the observed data, otherwise the test may be less powerful.

Values of D equal to or greater than that tabulated for a given pair of sample sizes n_1 and n_2 and for a specified significance level α suggest that the two samples have not come from the same population. The tabulated values of α relate to a two-sided alternative hypothesis regarding the difference between the sampled populations.

For values of n_1 and n_2 greater than 25, it is sufficiently informative in relation to *statistical* significance to calculate the critical values from the formulae given at the foot of the table.

n_1	α	n_2 2	3	4	5	6	7	8	9	10	11	12	13
2	0.05	–	–	–	–	–	–	1.000	1.000	1.000	1.000	1.000	1.000
2	0.025	–	–	–	–	–	–	–	–	–	–	1.000	1.000
2	0.01	–	–	–	–	–	–	–	–	–	–	–	–
2	0.001	–	–	–	–	–	–	–	–	–	–	–	–
3	0.05	–	–	–	1.000	1.000	1.000	.875	.889	.900	.909	.833	.846
3	0.025	–	–	–	–	1.000	1.000	1.000	1.000	1.000	.909	.917	.923
3	0.01	–	–	–	–	–	–	–	1.000	1.000	1.000	1.000	1.000
3	0.001	–	–	–	–	–	–	–	–	–	–	–	–
4	0.05	–	–	1.000	1.000	.833	.857	.875	.778	.750	.750	.750	.750
4	0.025	–	–	–	1.000	1.000	1.000	.875	.889	.900	.818	.833	.846
4	0.01	–	–	–	–	1.000	1.000	1.000	1.000	.900	.909	.917	.923
4	0.001	–	–	–	–	–	–	–	–	–	–	–	1.000
5	0.05	–	1.000	1.000	1.000	.800	.800	.750	.778	.800	.709	.717	.692
5	0.025	–	–	1.000	1.000	1.000	.857	.800	.800	.800	.800	.750	.723
5	0.01	–	–	–	1.000	1.000	1.000	.875	.889	.900	.818	.833	.800
5	0.001	–	–	–	–	–	–	–	1.000	1.000	1.000	1.000	1.000
6	0.05	–	1.000	.833	.800	.833	.714	.708	.722	.667	.652	.667	.667
6	0.025	–	1.000	1.000	1.000	1.000	.833	.750	.778	.733	.727	.750	.692
6	0.01	–	–	1.000	1.000	1.000	.857	.833	.833	.800	.818	.833	.769
6	0.001	–	–	–	–	–	–	1.000	1.000	1.000	1.000	.917	.923

Table 18 Percentage Points of D in the Two-sample Kolmogorov-Smirnov Distribution – continued

					n_2								
14	15	16	17	18	19	20	21	22	23	24	25	α	n_1
.929	.933	.938	.941	.944	.947	.950	.905	.909	.913	.917	.920	0.05	2
1.000	1.000	1.000	1.000	1.000	1.000	1.000	.952	.955	.957	.958	.960	0.025	2
–	–	–	–	–	1.000	1.000	1.000	1.000	1.000	1.000	1.000	0.01	2
–	–	–	–	–	–	–	–	–	–	–	–	0.001	2
.857	.800	.813	.824	.833	.789	.800	.810	.773	.783	.792	.800	0.05	3
.929	.867	.875	.882	.889	.895	.850	.857	.864	.870	.833	.840	0.025	3
1.000	.933	.938	.941	.944	.947	.950	.905	.909	.913	.917	.920	0.01	3
–	–	–	–	–	–	–	1.000	1.000	1.000	1.000	1.000	0.001	3
.750	.733	.750	.706	.694	.697	.750	.702	.705	.696	.708	.680	0.05	4
.786	.750	.813	.765	.750	.750	.800	.750	.750	.750	.750	.750	0.025	4
.857	.867	.875	.882	.833	.842	.850	.857	.818	.826	.833	.840	0.01	4
1.000	1.000	1.000	1.000	1.000	1.000	.950	.952	.955	.957	.958	.960	0.001	4
.657	.733	.675	.647	.667	.642	.650	.657	.636	.626	.633	.640	0.05	5
.729	.733	.738	.706	.722	.695	.750	.705	.709	.696	.675	.720	0.025	5
.800	.800	.800	.800	.778	.747	.800	.762	.755	.757	.750	.760	0.01	5
1.000	.933	.938	.941	.944	.895	.900	.905	.909	.913	.875	.880	0.001	5
.643	.633	.625	.608	.667	.614	.600	.595	.591	.580	.625	.587	0.05	6
.690	.700	.667	.657	.722	.667	.650	.643	.652	.623	.667	.640	0.025	6
.762	.767	.750	.716	.778	.728	.733	.714	.697	.703	.708	.713	0.01	6
.929	.933	.875	.833	.889	.842	.833	.833	.833	.826	.833	.833	0.001	6

Table 18 Percentage Points of D in the Two-sample Kolmogorov-Smirnov Distribution – continued

n_1	α	2	3	4	5	6	7	8	9	10	11	12	13
7	0.05	–	1.000	.857	.800	.714	.857	.714	.667	.657	.623	.631	.615
7	0.025	–	1.000	1.000	.857	.833	.857	.732	.714	.700	.675	.667	.637
7	0.01	–	–	1.000	1.000	.857	.857	.857	.778	.757	.766	.714	.714
7	0.001	–	–	–	–	–	1.000	1.000	1.000	.900	.909	.857	.857
8	0.05	1.000	.875	.875	.750	.708	.714	.750	.639	.600	.602	.625	.596
8	0.025	–	1.000	.875	.800	.750	.732	.750	.667	.675	.659	.667	.625
8	0.01	–	–	1.000	.875	.833	.857	.875	.764	.750	.727	.708	.692
8	0.001	–	–	–	–	1.000	1.000	1.000	.889	.875	.875	.833	.846
9	0.05	1.000	.889	.778	.778	.722	.667	.639	.667	.589	.596	.583	.556
9	0.025	–	1.000	.889	.800	.778	.714	.667	.778	.667	.636	.639	.615
9	0.01	–	1.000	1.000	.889	.833	.778	.764	.778	.700	.707	.694	.667
9	0.001	–	–	–	1.000	1.000	1.000	.889	.889	.889	.818	.806	.778
10	0.05	1.000	.900	.750	.800	.667	.657	.600	.589	.700	.545	.550	.538
10	0.025	–	1.000	.900	.800	.733	.700	.675	.667	.700	.618	.600	.592
10	0.01	–	1.000	.900	.900	.800	.757	.750	.700	.800	.700	.667	.646
10	0.001	–	–	–	1.000	1.000	.900	.875	.889	.900	.809	.800	.769
11	0.05	1.000	.909	.750	.709	.652	.623	.602	.596	.545	.636	.545	.524
11	0.025	–	.909	.818	.800	.727	.675	.659	.636	.618	.636	.576	.587
11	0.01	–	1.000	.909	.818	.818	.766	.727	.707	.700	.727	.652	.636
11	0.001	–	–	–	1.000	1.000	.909	.875	.818	.809	.818	.750	.755
12	0.05	1.000	.833	.750	.717	.667	.631	.625	.583	.550	.545	.583	.519
12	0.025	1.000	.917	.833	.750	.750	.667	.667	.639	.600	.576	.667	.538
12	0.01	–	1.000	.917	.833	.833	.714	.708	.694	.667	.652	.667	.609
12	0.001	–	–	–	1.000	.917	.857	.833	.806	.800	.750	.833	.750
13	0.05	1.000	.846	.750	.692	.667	.615	.596	.556	.538	.524	.519	.538
13	0.025	1.000	.923	.846	.723	.692	.637	.625	.615	.592	.587	.538	.615
13	0.01	–	1.000	.923	.800	.769	.714	.692	.667	.646	.636	.609	.692
13	0.001	–	–	1.000	1.000	.923	.857	.846	.778	.769	.755	.750	.769
14	0.05	.929	.857	.750	.657	.643	.643	.571	.556	.529	.532	.512	.489
14	0.025	1.000	.929	.786	.729	.690	.714	.625	.603	.586	.565	.560	.549
14	0.01	–	1.000	.857	.800	.762	.786	.679	.667	.643	.623	.619	.571
14	0.001	–	–	1.000	1.000	.929	.857	.804	.778	.757	.747	.714	.709
15	0.05	.933	.800	.733	.733	.633	.590	.558	.556	.533	.509	.517	.492
15	0.025	1.000	.867	.750	.733	.700	.648	.617	.600	.600	.570	.550	.533
15	0.01	–	.933	.867	.800	.767	.714	.675	.667	.667	.618	.600	.590
15	0.001	–	–	1.000	.933	.933	.857	.808	.778	.767	.727	.717	.703
16	0.05	.938	.813	.750	.675	.625	.571	.625	.542	.525	.506	.500	.486
16	0.025	1.000	.875	.813	.738	.667	.652	.625	.590	.563	.545	.542	.534
16	0.01	–	.938	.875	.800	.750	.688	.688	.653	.625	.602	.604	.582
16	0.001	–	–	1.000	.938	.875	.857	.813	.764	.738	.722	.708	.688
17	0.05	.941	.824	.706	.647	.608	.571	.566	.536	.524	.497	.490	.475
17	0.025	1.000	.882	.765	.706	.657	.647	.588	.588	.565	.545	.529	.516
17	0.01	–	.941	.882	.800	.716	.706	.647	.647	.624	.588	.583	.575
17	0.001	–	–	1.000	.941	.833	.824	.816	.765	.741	.706	.691	.688
18	0.05	.944	.833	.694	.667	.667	.571	.556	.556	.511	.490	.500	.470
18	0.025	1.000	.889	.750	.722	.722	.635	.597	.611	.556	.540	.556	.513
18	0.01	–	.944	.833	.778	.778	.690	.653	.667	.600	.596	.583	.560
18	0.001	–	–	1.000	.944	.889	.802	.778	.778	.733	.707	.694	.667

Table 18 Percentage Points of D in the Two-sample Kolmogorov-Smirnov Distribution – continued

					n_2									
14	15	16	17	18	19	20	21	22	23	24	25	α	n_1
.643	.590	.571	.571	.571	.571	.564	.619	.545	.553	.548	.554	0.05	7
.714	.648	.652	.647	.635	.632	.614	.667	.623	.609	.607	.600	0.025	7
.786	.714	.688	.706	.690	.684	.664	.714	.669	.671	.667	.657	0.01	7
.857	.857	.857	.824	.802	.805	.800	.810	.812	.783	.792	.777	0.001	7
.571	.558	.625	.566	.556	.539	.550	.530	.534	.533	.542	.520	0.05	8
.625	.617	.625	.588	.597	.592	.600	.577	.580	.576	.583	.560	0.025	8
.679	.675	.688	.647	.653	.645	.650	.637	.636	.625	.667	.625	0.01	8
.804	.808	.813	.816	.778	.770	.775	.750	.750	.745	.792	.750	0.001	8
.556	.556	.542	.536	.556	.520	.517	.524	.510	.512	.514	.507	0.05	9
.603	.600	.590	.588	.611	.573	.556	.571	.556	.556	.556	.547	0.025	9
.667	.667	.653	.647	.667	.626	.617	.619	.616	.609	.611	.600	0.01	9
.778	.778	.764	.765	.778	.737	.739	.730	.727	.734	.722	.720	0.001	9
.529	.533	.525	.524	.511	.495	.550	.500	.491	.496	.492	.500	0.05	10
.586	.600	.563	.565	.556	.542	.600	.552	.536	.539	.533	.540	0.025	10
.643	.667	.625	.624	.600	.595	.650	.600	.591	.596	.583	.600	0.01	10
.757	.767	.738	.741	.733	.700	.750	.710	.700	.696	.692	.700	0.001	10
.532	.509	.506	.497	.490	.488	.486	.485	.500	.470	.470	.469	0.05	11
.565	.570	.545	.545	.540	.531	.527	.532	.545	.518	.519	.509	0.025	11
.623	.618	.602	.588	.596	.584	.577	.580	.591	.561	.568	.560	0.01	11
.747	.727	.722	.706	.707	.699	.700	.680	.727	.684	.667	.669	0.001	11
.512	.517	.500	.490	.500	.474	.483	.476	.470	.453	.500	.460	0.05	12
.560	.550	.542	.529	.556	.526	.517	.512	.508	.496	.542	.500	0.025	12
.619	.600	.604	.583	.583	.570	.583	.560	.561	.540	.583	.550	0.01	12
.714	.717	.708	.691	.694	.684	.683	.667	.659	.659	.667	.640	0.001	12
.489	.492	.486	.475	.470	.462	.462	.462	.455	.452	.449	.446	0.05	13
.549	.533	.534	.516	.513	.510	.500	.502	.493	.488	.484	.486	0.025	13
.571	.590	.582	.575	.560	.559	.550	.549	.545	.538	.532	.529	0.01	13
.709	.703	.688	.688	.667	.664	.650	.656	.647	.639	.638	.615	0.001	13
.571	.467	.473	.466	.460	.455	.450	.476	.448	.441	.435	.429	0.05	14
.571	.524	.518	.513	.500	.500	.493	.500	.481	.478	.476	.474	0.025	14
.643	.586	.563	.563	.556	.556	.543	.548	.532	.528	.524	.520	0.01	14
.786	.667	.679	.668	.659	.662	.643	.643	.636	.627	.625	.626	0.001	14
.467	.533	.475	.455	.456	.446	.450	.438	.436	.432	.433	.427	0.05	15
.524	.600	.496	.506	.500	.495	.500	.486	.467	.472	.467	.467	0.025	15
.586	.600	.554	.557	.544	.533	.533	.533	.524	.519	.517	.520	0.01	15
.667	.733	.675	.647	.644	.632	.650	.629	.621	.609	.617	.613	0.001	15
.473	.475	.500	.456	.444	.438	.438	.432	.426	.427	.438	.418	0.05	16
.518	.496	.563	.500	.486	.477	.488	.467	.466	.459	.479	.453	0.025	16
.563	.554	.625	.526	.535	.526	.525	.515	.511	.508	.521	.498	0.01	16
.679	.675	.688	.640	.646	.632	.625	.619	.614	.601	.604	.595	0.001	16
.466	.455	.456	.471	.435	.437	.429	.423	.420	.417	.412	.407	0.05	17
.513	.506	.500	.529	.484	.467	.471	.465	.455	.458	.449	.447	0.025	17
.563	.557	.526	.588	.536	.514	.515	.504	.500	.501	.498	.487	0.01	17
.668	.647	.640	.706	.611	.619	.615	.608	.602	.593	.588	.586	0.001	17
.460	.456	.444	.435	.500	.415	.422	.421	.414	.411	.417	.400	0.05	18
.500	.500	.486	.484	.500	.465	.461	.460	.449	.444	.458	.436	0.025	18
.556	.544	.535	.536	.556	.515	.506	.500	.495	.493	.500	.480	0.01	18
.659	.644	.646	.611	.667	.620	.594	.595	.591	.585	.583	.571	0.001	18

Table 18 Percentage Points of D in the Two-sample Kolmogorov-Smirnov Distribution – continued

n_1	α	2	3	4	5	6	7	8	9	10	11	12	13
19	0.05	.947	.789	.697	.642	.614	.571	.539	.520	.495	.488	.474	.462
19	0.025	1.000	.895	.750	.695	.667	.632	.592	.573	.542	.531	.526	.510
19	0.01	1.000	.947	.842	.747	.728	.684	.645	.626	.595	.584	.570	.559
19	0.001	–	–	1.000	.895	.842	.805	.770	.737	.700	.699	.684	.664
20	0.05	.950	.800	.750	.650	.600	.564	.550	.517	.550	.486	.483	.462
20	0.025	1.000	.850	.800	.750	.650	.614	.600	.556	.600	.527	.517	.500
20	0.01	1.000	.950	.850	.800	.733	.664	.650	.617	.650	.577	.583	.550
20	0.001	–	–	.950	.900	.833	.800	.775	.739	.750	.700	.683	.650
21	0.05	.905	.810	.702	.657	.595	.619	.530	.524	.500	.485	.476	.462
21	0.025	.952	.857	.750	.705	.643	.667	.577	.571	.552	.532	.512	.502
21	0.01	1.000	.905	.857	.762	.714	.714	.637	.619	.600	.580	.560	.549
21	0.001	–	1.000	.952	.905	.833	.810	.750	.730	.710	.680	.667	.656
22	0.05	.909	.773	.705	.636	.591	.545	.534	.510	.491	.500	.470	.455
22	0.025	.955	.864	.750	.709	.652	.623	.580	.556	.536	.545	.508	.493
22	0.01	1.000	.909	.818	.755	.697	.669	.636	.616	.591	.591	.561	.545
22	0.001	–	1.000	.955	.909	.833	.812	.750	.727	.700	.727	.659	.647
23	0.05	.913	.783	.696	.626	.580	.553	.533	.512	.496	.470	.453	.452
23	0.025	.957	.870	.750	.696	.623	.609	.576	.556	.539	.518	.496	.488
23	0.01	1.000	.913	.826	.757	.703	.671	.625	.609	.596	.561	.540	.538
23	0.001	–	1.000	.957	.913	.826	.783	.745	.734	.696	.684	.659	.639
24	0.05	.917	.792	.708	.633	.625	.548	.542	.514	.492	.470	.500	.449
24	0.025	.958	.833	.750	.675	.667	.607	.583	.556	.533	.519	.542	.484
24	0.01	1.000	.917	.833	.750	.708	.667	.667	.611	.583	.568	.583	.532
24	0.001	–	1.000	.958	.875	.833	.792	.792	.722	.692	.667	.667	.638
25	0.05	.920	.800	.680	.640	.587	.554	.520	.507	.500	.469	.460	.446
25	0.025	.960	.840	.750	.720	.640	.600	.560	.547	.540	.509	.500	.486
25	0.01	1.000	.920	.840	.760	.713	.657	.625	.600	.600	.560	.550	.529
25	0.001	–	1.000	.960	.880	.833	.777	.750	.720	.700	.669	.640	.615

α	0.05	0.025	0.01	0.001
D_α	$1.36\sqrt{\dfrac{n_1+n_2}{n_1 n_2}}$	$1.48\sqrt{\dfrac{n_1+n_2}{n_1 n_2}}$	$1.63\sqrt{\dfrac{n_1+n_2}{n_1 n_2}}$	$1.95\sqrt{\dfrac{n_1+n_2}{n_1 n_2}}$

Table 18 Percentage Points of D in the Two-sample Kolmogorov-Smirnov Distribution – continued

					n_2								
14	15	16	17	18	19	20	21	22	23	24	25	α	n_1
.455	.446	.438	.437	.415	.474	.421	.409	.404	.405	.401	.394	0.05	19
.500	.495	.477	.467	.465	.526	.445	.451	.443	.435	.436	.432	0.025	19
.556	.533	.526	.514	.515	.526	.492	.499	.488	.478	.478	.472	0.01	19
.662	.632	.632	.619	.620	.632	.592	.594	.579	.579	.572	.564	0.001	19
.450	.450	.438	.429	.422	.421	.450	.412	.400	.400	.400	.400	0.05	20
.493	.500	.488	.471	.461	.445	.500	.429	.436	.433	.433	.430	0.025	20
.543	.533	.525	.515	.506	.492	.550	.474	.482	.476	.475	.470	0.01	20
.643	.650	.625	.615	.594	.592	.650	.569	.577	.570	.567	.560	0.001	20
.476	.438	.432	.423	.421	.409	.412	.429	.396	.391	.393	.385	0.05	21
.500	.486	.467	.465	.460	.451	.429	.476	.439	.427	.423	.419	0.025	21
.548	.533	.515	.504	.500	.499	.474	.524	.483	.470	.470	.465	0.01	21
.643	.629	.619	.608	.595	.594	.569	.619	.578	.557	.560	.552	0.001	21
.448	.436	.426	.420	.414	.404	.400	.396	.409	.383	.386	.380	0.05	22
.481	.467	.466	.455	.449	.443	.436	.439	.455	.423	.420	.415	0.025	22
.532	.524	.511	.500	.495	.488	.482	.483	.500	.468	.458	.455	0.01	22
.636	.621	.614	.602	.591	.579	.577	.578	.591	.557	.553	.544	0.001	22
.441	.432	.427	.417	.411	.405	.400	.391	.383	.435	.371	.376	0.05	23
.478	.472	.459	.458	.444	.435	.433	.427	.423	.435	.409	.412	0.025	23
.528	.519	.508	.501	.493	.478	.476	.470	.468	.478	.451	.456	0.01	23
.627	.609	.601	.593	.585	.579	.570	.557	.557	.565	.536	.543	0.001	23
.435	.433	.438	.412	.417	.401	.400	.393	.386	.371	.417	.375	0.05	24
.476	.467	.479	.449	.458	.436	.433	.423	.420	.409	.458	.397	0.025	24
.524	.517	.521	.498	.500	.478	.475	.470	.458	.451	.500	.437	0.01	24
.625	.617	.604	.588	.583	.572	.567	.560	.553	.536	.583	.520	0.001	24
.429	.427	.418	.407	.400	.394	.400	.385	.380	.376	.375	.400	0.05	25
.474	.467	.453	.447	.436	.432	.430	.419	.415	.412	.397	.440	0.025	25
.520	.520	.498	.487	.480	.472	.470	.465	.455	.456	.437	.480	0.01	25
.626	.613	.595	.586	.571	.564	.560	.552	.544	.543	.520	.560	0.001	25

Table 19 Lower Percentage Points of the Mann-Whitney U-Distribution

The table gives the critical values of the Mann-Whitney statistic U for one-tailed tests for $\alpha = 0.05, 0.025, 0.01$ and 0.001. For two-tailed tests the probabilities should be doubled.

Given two random samples of size n_1 and n_2 $(n_1 \leq n_2)$, rank the $n_1 + n_2$ observations in increasing order of size. If there are ties, give each tied rank the average of the ranks they would each have had if there were no ties.

Denote the sum of the ranks of the sample of size n_1 by R_1 and that obtained from n_2 by R_2.

Calculate $u_1 = n_1 n_2 + \frac{1}{2} n_1 (n_1 + 1) - R_1$

and $\qquad u_2 = n_1 n_2 + \frac{1}{2} n_2 (n_2 + 1) - R_2$ although since $u_1 + u_2 = n_1 n_2$, this can more easily be obtained as $n_1 n_2 - u_1$.

For an alternative hypothesis $\mu_1 < \mu_2$, the observed value, u_1, should be less than the tabulated value of U in order to reject the hypothesis that $\mu_1 = \mu_2$ at the one-sided α level of significance.

To reject equality of population means in favour of the one-sided alternative $\mu_1 > \mu_2$, u_2 should be less than U.

For the two-sided alternative $\mu_1 \neq \mu_2$, reject the equality of means hypothesis if the smaller of u_1 and u_2 is less than the tabulated value of U though the significance level of this test will be double the quoted value of α.

Note that the quoted significance levels, α, are not generally exact, especially for small n_1 and n_2. The exact Type I error, in all cases, will not exceed the quoted value of α, being the highest feasible value less than α. For example, for $n_1 = 4$ and $n_2 = 6$, the exact probability of observing a U-value of 3 or less is 0.033, not 0.05. The tabulation is thus conservative regarding Type I errors.

n_1	α	3	4	5	6	7	8	9	10	11	12	13	14	15	16	17	18	19	20
1	0.05	–	–	–	–	–	–	–	–	–	–	–	–	–	–	–	–	0	0
1	0.025	–	–	–	–	–	–	–	–	–	–	–	–	–	–	–	–	–	–
1	0.01	–	–	–	–	–	–	–	–	–	–	–	–	–	–	–	–	–	–
1	0.001	–	–	–	–	–	–	–	–	–	–	–	–	–	–	–	–	–	–
2	0.05	–	–	0	0	0	1	1	1	1	2	2	3	3	3	3	4	4	4
2	0.025	–	–	–	–	–	0	0	0	0	1	1	1	1	1	2	2	2	2
2	0.01	–	–	–	–	–	–	–	–	–	0	0	0	0	0	0	0	1	1
2	0.001	–	–	–	–	–	–	–	–	–	–	–	–	–	–	–	–	–	–
3	0.05	0	0	1	2	2	3	4	4	5	5	6	7	7	8	9	9	10	11
3	0.025	–	–	0	1	1	2	2	3	3	4	4	5	5	6	6	7	7	8
3	0.01	–	–	–	–	0	0	1	1	1	2	2	2	3	3	4	4	4	5
3	0.001	–	–	–	–	–	–	–	–	–	–	–	–	–	0	0	0	0	0
4	0.05	–	1	2	3	4	5	6	7	8	9	10	11	12	14	15	16	17	18
4	0.025	–	0	1	2	3	4	4	5	6	7	8	9	10	11	11	12	13	13
4	0.01	–	–	0	1	1	2	3	3	4	5	5	6	7	7	8	9	9	10
4	0.001	–	–	–	–	–	–	–	0	0	0	1	1	1	2	2	3	3	3
5	0.05	–	–	4	5	6	8	9	11	12	13	15	16	18	19	20	22	23	25
5	0.025	–	–	2	3	5	6	7	8	9	11	12	13	14	15	17	18	19	20
5	0.01	–	–	1	2	3	4	5	6	7	8	9	10	11	12	13	14	15	16
5	0.001	–	–	–	–	0	0	1	1	2	2	3	3	4	5	5	6	7	7
6	0.05	–	–	–	7	8	10	12	14	16	17	19	21	23	25	26	28	30	32
6	0.025	–	–	–	5	6	8	10	11	13	14	16	17	19	21	22	24	25	27
6	0.01	–	–	–	3	4	6	7	8	9	11	12	13	15	16	18	19	20	22
6	0.001	–	–	–	0	1	2	2	3	4	4	5	6	7	8	9	10	11	12
7	0.05	–	–	–	–	11	13	15	17	19	21	24	26	28	30	33	35	37	39
7	0.025	–	–	–	–	8	10	12	14	16	18	20	22	24	26	28	30	32	34
7	0.01	–	–	–	–	6	8	9	11	12	14	16	17	19	21	23	24	26	28
7	0.001	–	–	–	–	2	3	3	5	6	7	8	9	10	11	13	14	15	16
8	0.05	–	–	–	–	–	15	18	20	23	26	28	31	33	36	39	41	44	47
8	0.025	–	–	–	–	–	13	15	17	19	22	24	26	29	31	34	36	38	41
8	0.01	–	–	–	–	–	10	11	13	15	17	20	22	24	26	28	30	32	34
8	0.001	–	–	–	–	–	5	5	6	8	9	11	12	14	15	17	18	20	21
9	0.05	–	–	–	–	–	–	21	24	27	30	33	36	39	42	45	48	51	54
9	0.025	–	–	–	–	–	–	17	20	23	26	28	31	34	37	39	42	45	48
9	0.01	–	–	–	–	–	–	14	16	18	21	23	26	28	31	33	36	38	40
9	0.001	–	–	–	–	–	–	7	8	10	12	14	15	17	19	21	23	25	26

Table 19 Lower Percentage Points of the Mann-Whitney U-Distribution

n_1	α	3	4	5	6	7	8	9	10	11	12	13	14	15	16	17	18	19	20
10	0.05	–	–	–	–	–	–	–	27	31	34	37	41	44	48	51	55	58	62
10	0.025	–	–	–	–	–	–	–	23	26	29	33	36	39	42	45	48	52	55
10	0.01	–	–	–	–	–	–	–	19	22	24	27	30	33	36	38	41	44	47
10	0.001	–	–	–	–	–	–	–	10	12	14	17	19	21	23	25	27	29	32
11	0.05	–	–	–	–	–	–	–	–	34	38	42	46	50	54	57	61	65	69
11	0.025	–	–	–	–	–	–	–	–	30	33	37	40	44	47	51	55	58	62
11	0.01	–	–	–	–	–	–	–	–	25	28	31	34	37	41	44	47	50	53
11	0.001	–	–	–	–	–	–	–	–	15	17	20	22	24	27	29	32	34	37
12	0.05	–	–	–	–	–	–	–	–	–	42	47	51	55	60	64	68	72	77
12	0.025	–	–	–	–	–	–	–	–	–	37	41	45	49	53	57	61	65	69
12	0.01	–	–	–	–	–	–	–	–	–	31	35	38	42	46	49	53	56	60
12	0.001	–	–	–	–	–	–	–	–	–	20	23	25	28	31	34	37	40	42
13	0.05	–	–	–	–	–	–	–	–	–	–	51	56	61	65	70	75	80	84
13	0.025	–	–	–	–	–	–	–	–	–	–	45	50	54	59	63	67	72	76
13	0.01	–	–	–	–	–	–	–	–	–	–	39	43	47	51	55	59	63	67
13	0.001	–	–	–	–	–	–	–	–	–	–	26	29	32	35	38	42	45	48
14	0.05	–	–	–	–	–	–	–	–	–	–	–	61	66	71	77	82	87	92
14	0.025	–	–	–	–	–	–	–	–	–	–	–	55	59	64	67	74	78	83
14	0.01	–	–	–	–	–	–	–	–	–	–	–	47	51	56	60	65	69	73
14	0.001	–	–	–	–	–	–	–	–	–	–	–	32	36	39	43	46	50	54
15	0.05	–	–	–	–	–	–	–	–	–	–	–	–	72	77	83	88	94	100
15	0.025	–	–	–	–	–	–	–	–	–	–	–	–	64	70	75	80	85	90
15	0.01	–	–	–	–	–	–	–	–	–	–	–	–	56	61	66	70	75	80
15	0.001	–	–	–	–	–	–	–	–	–	–	–	–	40	43	47	51	55	59
16	0.05	–	–	–	–	–	–	–	–	–	–	–	–	–	83	89	95	101	107
16	0.025	–	–	–	–	–	–	–	–	–	–	–	–	–	75	81	86	92	98
16	0.01	–	–	–	–	–	–	–	–	–	–	–	–	–	66	71	76	82	87
16	0.001	–	–	–	–	–	–	–	–	–	–	–	–	–	48	52	56	60	65
17	0.05	–	–	–	–	–	–	–	–	–	–	–	–	–	–	96	102	109	115
17	0.025	–	–	–	–	–	–	–	–	–	–	–	–	–	–	87	93	99	105
17	0.01	–	–	–	–	–	–	–	–	–	–	–	–	–	–	77	82	88	93
17	0.001	–	–	–	–	–	–	–	–	–	–	–	–	–	–	57	61	66	70
18	0.05	–	–	–	–	–	–	–	–	–	–	–	–	–	–	–	109	116	123
18	0.025	–	–	–	–	–	–	–	–	–	–	–	–	–	–	–	99	106	112
18	0.01	–	–	–	–	–	–	–	–	–	–	–	–	–	–	–	88	94	100
18	0.001	–	–	–	–	–	–	–	–	–	–	–	–	–	–	–	66	71	76
19	0.05	–	–	–	–	–	–	–	–	–	–	–	–	–	–	–	–	123	130
19	0.025	–	–	–	–	–	–	–	–	–	–	–	–	–	–	–	–	113	119
19	0.01	–	–	–	–	–	–	–	–	–	–	–	–	–	–	–	–	101	107
19	0.001	–	–	–	–	–	–	–	–	–	–	–	–	–	–	–	–	77	82
20	0.05	–	–	–	–	–	–	–	–	–	–	–	–	–	–	–	–	–	138
20	0.025	–	–	–	–	–	–	–	–	–	–	–	–	–	–	–	–	–	127
20	0.01	–	–	–	–	–	–	–	–	–	–	–	–	–	–	–	–	–	114
20	0.001	–	–	–	–	–	–	–	–	–	–	–	–	–	–	–	–	–	88

For values of $n_2 > 20$, the sampling distribution of U rapidly tends to normality with mean $= \frac{1}{2}n_1 n_2$

and standard deviation $= \sqrt{\dfrac{n_1 n_2 (n_1 + n_2 + 1)}{12}}$

Thus we have that the statistic $\dfrac{U - \frac{1}{2} n_1 n_2}{\sqrt{\dfrac{n_1 n_2 (n_1 + n_2 + 1)}{12}}}$ will be distributed like the unit normal variate, Z.

37

Table 20 Percentage Points of Friedman's Distribution

The table gives the probabilities associated with values as large as, or larger than, the possible observed values of Friedman's statistic χ_r^2 (also denoted in other sources by M).

    Given k matched samples, each therefore of size n, and with the observations lying on at least an ordinal scale, Friedman's distribution may be used to test the hypothesis that the k groups have a common mean value. This is equivalent to a standard balanced analysis of variance but without the requirement that each population is normally distributed.

    The data are laid out in n rows (equivalent to blocks) and k columns and the observations in each row are converted to their equivalent ranks between 1 and k. If there are ties, the appropriate average rank is used. The total, R_j, of the ranks in each column is used to form the sum of these squared rank totals.

Friedman's statistic is calculated from the expression

$$\chi_r^2 = \frac{12}{nk(k+1)} \sum_{j=1}^{k} R_j^2 - 3n(k+1)$$

$k = 3$

| $n=2$ | | $n=3$ | | $n=4$ | | $n=5$ | |
χ_r^2	p	χ_r^2	p	χ_r^2	p	χ_r^2	p
0	1.000	.000	1.000	.0	1.000	.0	1.000
1	.833	.667	.944	.5	.931	.4	.954
3	.5	2.000	.528	1.5	.653	1.2	.691
4	.167	2.667	.361	2.0	.431	1.6	.522
-	-	4.667	.194	3.5	.273	2.8	.367
-	-	6.000	.028	4.5	.125	3.6	.182
-	-	-	-	6.0	.069	4.8	.124
-	-	-	-	6.5	.042	5.2	.093
-	-	-	-	8.0	.0046	6.4	.039
-	-	-	-	-	-	7.6	.024
-	-	-	-	-	-	8.4	.0085
-	-	-	-	-	-	10.0	.00077

$k = 3$

| $n=6$ | | $n=7$ | | $n=8$ | | $n=9$ | |
χ_r^2	p	χ_r^2	p	χ_r^2	p	χ_r^2	p
.00	1.000	.000	1.000	.00	1.000	.000	1.000
.33	.956	.286	.964	.25	.967	.222	.971
1.00	.740	.857	.768	.75	.794	.667	.814
1.33	.570	1.143	.620	1.00	.654	.889	.865
2.33	.430	2.000	.486	1.75	.531	1.556	.569
3.00	.252	2.571	.305	2.25	.355	2.000	.398
4.00	.184	3.429	.237	3.00	.285	2.667	.328
4.33	.142	3.714	.192	3.25	.236	2.889	.278
5.33	.072	4.571	.112	4.00	.149	3.556	.187
6.33	.052	5.429	.085	4.75	.120	4.222	.154
7.00	.029	6.000	.052	5.25	.079	4.667	.107
8.33	.012	7.143	.027	6.25	.047	5.556	.069
9.00	.0081	7.714	.021	6.75	.038	6.000	.057
9.33	.0055	8.000	.016	7.00	.030	6.222	.048
10.33	.0017	8.857	.0084	7.75	.018	6.889	.031
12.00	.00013	10.286	.0036	9.00	.0099	8.000	.019
		10.571	.0027	9.25	.0080	8.222	.016
		11.143	.0012	9.75	.0048	8.667	.010
		12.286	.00032	10.75	.0024	9.556	.0060
		14.000	.000021	12.00	.0011	10.667	.0035
				12.25	.00086	10.889	.0029
				13.00	.00026	11.556	.0013
				14.25	.000061	12.667	.00066
				16.00	.0000036	13.556	.00035
						14.000	.00020
						14.222	.000097
						14.889	.000054
						16.222	.000011
						18.000	.0000006

Table 20 Percentage Points of Friedman's Distribution – continued

$k = 4$

$n = 2$		$n = 3$		$n = 4$			
χ_r^2	p	χ_r^2	p	χ_r^2	p	χ_r^2	p
.0	1.000	.2	1.000	.0	1.000	5.7	.141
.6	.958	.6	.958	.3	.992	6.0	.105
1.2	.834	1.0	.910	.6	.928	6.3	.094
1.8	.792	1.8	.727	.9	.900	6.6	.077
2.4	.625	2.2	.608	1.2	.800	6.9	.068
3.0	.542	2.6	.524	1.5	.754	7.2	.054
3.6	.458	3.4	.446	1.8	.677	7.5	.052
4.2	.375	3.8	.342	2.1	.649	7.8	.036
4.8	.208	4.2	.300	2.4	.524	8.1	.033
5.4	.167	5.0	.207	2.7	.508	8.4	.019
6.0	.042	5.4	.175	3.0	.432	8.7	.014
		5.8	.148	3.3	.389	9.3	.012
		6.6	.075	3.6	.355	9.6	.0069
		7.0	.054	3.9	.324	9.9	.0062
		7.4	.033	4.5	.242	10.2	.0027
		8.2	.017	4.8	.200	10.8	.0016
		9.0	.0017	5.1	.190	11.1	.00094
				5.4	.158	12.0	.000072

For larger values of n and k than tabulated, the associated probability may be found from a table of the χ^2 distribution (e.g. Table 8) since Friedman's statistic χ_r^2 tends towards a χ^2-distribution with $(k-1)$ degrees of freedom as n and k increase.

Table 21 Upper Tails of the Kruskal-Wallis Distribution

Sample sizes					Sample sizes				
n_1	n_2	n_3	H	p	n_1	n_2	n_3	H	p
2	1	1	2.7000	.500	4	3	2	4.4444	.102
								4.5111	.098
2	2	1	3.6000	.200				5.4000	.051
								5.4444	.046
2	2	2	3.7143	.200				6.3000	.011
			4.5714	.067				6.4444	.008
3	1	1	3.2000	.300	4	3	3	4.7000	.101
								4.7091	.092
3	2	1	3.8571	.133				5.7273	.050
			4.2857	.100				5.7909	.046
								6.7091	.013
3	2	2	4.4643	.105				6.7455	.010
			4.5000	.067					
			4.7143	.048	4	4	1	4.0667	.102
			5.3572	.029				4.1667	.082
								4.8667	.054
3	3	1	4.0000	.129				4.9667	.048
			4.5714	.100				6.1667	.022
			5.1429	.043				6.6667	.010
3	3	2	4.2500	.121	4	4	2	4.4455	.103
			4.5556	.100				4.5545	.098
			5.1389	.061				5.2364	.052
			5.3611	.032				5.4545	.046
			6.2500	.011				6.8727	.011
								7.0364	.006
3	3	3	4.6222	.100					
			5.0667	.086	4	4	3	4.4773	.102
			5.6000	.050				4.5455	.099
			5.6889	.029				5.5758	.051
			6.4889	.011				5.5985	.049
			7.2000	.004				7.1364	.011
								7.1439	.010
4	1	1	3.5714	.200					
					4	4	4	4.5001	.104
4	2	1	4.0179	.114				4.6539	.097
			4.5000	.076				5.6538	.054
			4.8214	.057				5.6923	.049
								7.5385	.011
4	2	2	4.1667	.105				7.6538	.008
			4.4583	.100					
			5.1250	.052	5	1	1	3.8571	.143
			5.3333	.033					
			6.0000	.014	5	2	1	4.0500	.119
								4.2000	.095
4	3	1	3.8889	.129				4.4500	.071
			4.0556	.093				5.0000	.048
			5.0000	.057				5.2500	.036
			5.2083	.050					
			5.8333	.021					
5	2	2	4.2933	.122	5	4	4	4.5527	.102
			4.3733	.090				4.6187	.100
			5.0400	.056				5.6176	.050
			5.1600	.034				5.6571	.049
			6.1333	.013				7.7440	.011
			6.5333	.008				7.76.4	.009

Table 21 Upper Tails of the Kruskal-Wallis Distribution - continued

Sample sizes					Sample sizes				
n_1	n_2	n_3	H	p	n_1	n_2	n_3	H	p
5	3	1	3.8400	.123	5	5	1	4.0364	.105
			4.0178	.095				4.1091	.086
			4.8711	.052				4.9091	.053
			4.9600	.048				5.1273	.046
			6.4000	.012				6.8364	.011
								7.3091	.009
5	3	2	4.4945	.101					
			4.6509	.091	5	5	2	4.5077	.100
			5.1055	.052				4.6231	.097
			5.2509	.049				5.2462	.051
			6.8218	.010				5.3385	.047
			6.9091	.009				7.2692	.010
								7.3385	.010
5	3	3	4.4121	.109					
			4.5333	.097	5	5	3	4.5363	.102
			5.5152	.051				4.5451	.100
			5.6485	.049				5.6264	.051
			6.9818	.011				5.7055	.046
			7.0788	.009				7.5429	.010
								7.5780	.010
5	4	1	3.9600	.102					
			3.9873	.098	5	5	4	4.5200	.101
			4.8600	.056				4.5229	.099
			4.9855	.044				5.6429	.050
			6.8400	.011				5.6657	.049
			6.9545	.008				7.7914	.010
								7.8229	.010
5	4	2	4.5182	.101					
			4.5409	.098	5	5	5	4.5000	.102
			5.2682	.050				4.5600	.100
			5.2727	.049				5.6600	.051
			7.1182	.010				5.7800	.049
			7.2045	.009				7.9800	.010
								8.0000	.009
5	4	3	4.5231	.103					
			4.5487	.099					
			5.6308	.050					
			5.6564	.049					
			7.3949	.011					
			7.4449	.010					

Given a set of random samples, not necessarily of equal size, obtained from each of k populations, the hypothesis of equality of the population means may be tested. The measured variable should produce at least ordinal values so that the responses can be ranked without many ties. The structure of the test is that of a one-way analysis of variance but without the need to assume a normal distribution.

To apply the test, convert the whole data set to one sequence of ascending order of ranks. The smallest observation would thus have rank 1, the next larger one than this would be assigned the rank 2 and so on. It is convenient to imagine the ranks laid out in k columns corresponding to the original data.

The total number of readings
$$n = \sum_{j=1}^{k} n_j$$
where n_j is the number of observations (ranks) in column j.

The statistic H is calculated from
$$H = \frac{12}{n(n+1)} \sum_{j=1}^{k} \frac{R_j^2}{n_j} - 3(n+1)$$

The table shows for $k = 3$, the probability that, for group sample sizes of n_1, n_2 and n_3, the statistic H takes certain preferred values or higher ones in the upper tail of the distribution. Thus for sample sizes 4, 4 and 2, a value of $H = 5.2364$ or higher would occur with probability of 0.052 just by chance when there is no actual difference between the k population means. The next possible higher value of H is 5.4545 with probability 0.046.

As n tends to infinity, the distribution of H tends to that of χ^2 with $(k - 1)$ df. This can be used as a reasonable approximation for H, given values of n_j and k outside the tabulated range.

For example, for $n_1 = n_2 = n_3 = 5$, the upper 5% point of H is 5.78 if you want to be conservative or 5.66 if you do not. The 5% point of χ^2 with 2 df is 5.99 and thus neither of the possible actual values, 5.66 or 5.78, would therefore be judged significant at the 5% level using the approximation. However, the probabilities that χ^2 with 2 df exceeds 5.66 and 5.78 are 0.059 and 0.056 respectively; thus from the point of view of practical interpretation of p-values, there is very little to choose between the exact approach and the χ^2 approximation.

Table 22 Control Chart Factors for Sample Mean $(\overline{X})$

To obtain the limits,

EITHER multiply σ by the appropriate value of $A_{0.025}$ and $A_{0.001}$

OR   multiply $\overline{R}$ by the appropriate value of $A'_{0.025}$ and $A'_{0.001}$

then add the result to and subtract it from the centre line of the chart ($\overline{\overline{X}}$ or μ).

No. in sample	For inner limits $A_{0.025}$	For outer limits $A_{0.001}$	For inner limits $A'_{0.025}$	For outer limits $A'_{0.001}$
2	1.386	2.185	1.229	1.937
3	1.132	1.784	0.668	1.054
4	0.980	1.545	0.476	0.750
5	0.876	1.382	0.377	0.594
6	0.800	1.262	0.316	0.498
7	0.741	1.168	0.274	0.432
8	0.693	1.092	0.244	0.384
9	0.653	1.030	0.220	0.347
10	0.620	0.977	0.202	0.317
11	0.591	0.932	0.186	0.294
12	0.566	0.892	0.174	0.274
13	0.544	0.857		
14	0.524	0.826		
15	0.506	0.798		
16	0.490	0.773		
17	0.475	0.750		
18	0.462	0.728		
19	0.450	0.709		
20	0.438	0.691		
21	0.428	0.674		
22	0.418	0.659		
23	0.409	0.644		
24	0.400	0.631		
25	0.392	0.618		
26	0.384	0.606		
27	0.377	0.595		
28	0.370	0.584		
29	0.364	0.574		
30	0.358	0.564		

Samples containing more than 12 individuals should not be used when utilising the range as a substitute for the standard deviation in calculating the $\overline{X}$ control limits

Table 23 Control Chart Factors for Sample Range Using $\overline{R}$

To obtain the limits, multiply $\overline{R}$ by the appropriate value of D' (the subscript of each D' value is the probability of a sample range of n readings being *larger* than the calculated limit).

No. in sample n	For lower limits		For upper limits	
	Action $D'_{0.999}$	Warning $D'_{0.975}$	Warning $D'_{0.025}$	Action $D'_{0.001}$
2	0.00	0.04	2.81	4.12
3	0.04	0.18	2.17	2.98
4	0.10	0.29	1.93	2.57
5	0.16	0.37	1.81	2.34
6	0.21	0.42	1.72	2.21
7	0.26	0.46	1.66	2.11
8	0.29	0.50	1.62	2.04
9	0.32	0.52	1.58	1.99
10	0.35	0.54	1.56	1.93
11	0.38	0.56	1.53	1.91
12	0.40	0.58	1.51	1.87

Table 24 Control Chart Factors for Sample Range Using σ

To obtain the limits, multiply σ by the appropriate value of D (the subscript of each D value is the probability of a sample range of n readings being *larger* than the calculated limit).

To obtain the expected value of sample range $\left(\overline{R}\right)$ multiply σ by the appropriate value of d_2.

No. in sample n	For lower limits		For upper limits		
	Action $D_{0.999}$	Warning $D_{0.975}$	Warning $D_{0.025}$	Action $D_{0.001}$	d_2
2	0.00	0.04	3.17	4.65	1.128
3	0.06	0.30	3.68	5.05	1.693
4	0.20	0.59	3.98	5.30	2.059
5	0.37	0.85	4.20	5.45	2.326
6	0.54	1.06	4.36	5.60	2.534
7	0.69	1.25	4.49	5.70	2.704
8	0.83	1.41	4.61	5.80	2.847
9	0.96	1.55	4.70	5.90	2.970
10	1.08	1.67	4.79	5.95	3.078
11	1.20	1.78	4.86	6.05	3.173
12	1.30	1.88	4.92	6.10	3.258

Table 25 Control Chart Factors for Mean and Range (American Usage)

X-bar and R charts

Obser-vations in sample	AVERAGES	RANGES				
	Factors for Control Limits	Factors for Central Line		Factors for Control Limits		
n	A_2	d_2	$1/d_2$	d_3	D_3	D_4
2	1.880	1.128	0.8865	0.853	0	3.267
3	1.023	1.693	0.5907	0.888	0	2.574
4	0.729	2.059	0.4857	0.880	0	2.282
5	0.577	2.326	0.4299	0.864	0	2.114
6	0.483	2.534	0.3946	0.848	0	2.004
7	0.419	2.704	0.3698	0.833	0.076	1.924
8	0.373	2.847	0.3512	0.820	0.136	1.864
9	0.337	2.970	0.3367	0.808	0.184	1.816
10	0.308	3.078	0.3249	0.797	0.223	1.777
11	0.285	3.173	0.3152	0.787	0.256	1.744
12	0.266	3.258	0.3069	0.778	0.283	1.717
13	0.249	3.336	0.2998	0.770	0.307	1.693
14	0.235	3.407	0.2935	0.763	0.328	1.672
15	0.223	3.472	0.2880	0.756	0.347	1.653
16	0.212	3.532	0.2831	0.750	0.363	1.637
17	0.203	3.588	0.2787	0.744	0.378	1.622
18	0.194	3.640	0.2747	0.739	0.391	1.608
19	0.187	3.689	0.2711	0.734	0.403	1.597
20	0.180	3.735	0.2677	0.729	0.415	1.585
21	0.173	3.778	0.2647	0.724	0.425	1.575
22	0.167	3.819	0.2618	0.720	0.434	1.566
23	0.162	3.858	0.2592	0.716	0.443	1.557
24	0.157	3.895	0.2567	0.712	0.451	1.548
25	0.153	3.931	0.2544	0.708	0.459	1.541

To derive control limits

Process Average $\overline{X}$ Chart

Upper Control Limit = Central Line +A_2 $\overline{R}$
Lower Control Limit = Central Line –A_2 $\overline{R}$

Central Line = $\overline{X}$ for trial limits
Central Line = Target mean for a controlled process

Range Chart

Upper Control Limit = D_4 $\overline{R}$
Lower Control Limit = D_3 $\overline{R}$

Central Line = $\overline{R}$ for trial limits
For achievable known process σ, replace $\overline{R}$ by $d_2 \sigma$.

Table 26 Control Chart Factors for Standard Deviation (American Usage)

Trial Control Limits (μ and σ not known)

S-chart: Central Line $= \dfrac{1}{k}\sum_{i=1}^{k} S_i = \bar{S}$: Upper Control Limit (UCL) $= B_4\bar{S}$: Lower Control Limit (LCL) $= B_3\bar{S}$

X-bar chart: Central Line $= \dfrac{1}{k}\sum_{i=1}^{k} \bar{X} = \bar{\bar{X}}$: UCL $= \bar{\bar{X}} + A_3\bar{S}$: LCL $= \bar{\bar{X}} - A_3\bar{S}$ where $A_3 = \dfrac{3}{c_4\sqrt{n}}$

Known Process σ

The factors B_5 and B_6 are applicable in the (probably unusual) case where the process variability is known to be stable and a good prior estimate of its standard deviation σ is known.

S-chart: Central Line $= c_4\sigma$:   UCL $= B_6\sigma$:   LCL $= B_5\sigma$

The central line of the **X-bar chart** would be either at $\bar{\bar{X}}$ for trial limits or at the process target mean.

UCL = Central Line $+3\dfrac{\sigma}{\sqrt{n}}$ LCL = Central Line $-3\dfrac{\sigma}{\sqrt{n}}$

Obser-vations in Sample	AVERAGES	STANDARD DEVIATIONS						
	Control Limit Factors	Factors for Central Line			Factors for Control Limits			
n	A_3	c_4	$1/c_4$	B_3	B_4	B_5	B_6
2	2.659	0.7979	1.2533	0	3.267	0	2.606
3	1.954	0.8862	1.1284	0	2.568	0	2.276
4	1.628	0.9213	1.0854	0	2.266	0	2.088
5	1.427	0.9400	1.0638	0	2.089	0	1.964
6	1.287	0.9515	1.0510	0.030	1.970	0.029	1.874
7	1.182	0.9594	1.0423	0.118	1.882	0.113	1.806
8	1.099	0.9650	1.0363	0.185	1.815	0.179	1.751
9	1.032	0.9693	1.0317	0.239	1.761	0.232	1.707
10	0.975	0.9727	1.0281	0.284	1.716	0.276	1.669
11	0.927	0.9754	1.0252	0.321	1.679	0.313	1.637
12	0.886	0.9776	1.0229	0.354	1.646	0.346	1.610
13	0.850	0.9794	1.0210	0.382	1.618	0.374	1.585
14	0.817	0.9810	1.0194	0.406	1.594	0.399	1.563
15	0.789	0.9823	1.0180	0.428	1.572	0.421	1.544
16	0.763	0.9835	1.0168	0.448	1.552	0.440	1.526
17	0.739	0.9845	1.0157	0.466	1.534	0.458	1.511
18	0.718	0.9854	1.0148	0.482	1.518	0.475	1.496
19	0.698	0.9862	1.0140	0.497	1.503	0.490	1.483
20	0.680	0.9869	1.0133	0.510	1.490	0.504	1.470
21	0.663	0.9876	1.0126	0.523	1.477	0.516	1.459
22	0.647	0.9882	1.0119	0.534	1.466	0.528	1.448
23	0.633	0.9887	1.0114	0.545	1.455	0.539	1.438
24	0.619	0.9892	1.0109	0.555	1.445	0.549	1.429
25	0.606	0.9896	1.0105	0.565	1.435	0.559	1.420

The above factors assume that k random samples each of size n are available from a normal distribution of a variable, X, with mean μ and variance σ^2.

For each sample, $\bar{X} = \dfrac{\sum X}{n}$ and $S = \sqrt{\dfrac{\sum (X - \bar{X})^2}{n-1}}$, both summations being taken over n.

Then $E[\bar{X}] = \mu$; $\mathrm{Var}[\bar{X}] = \sigma^2/n$; $E[S] = c_4\sigma$; $\mathrm{Var}[S] = \sigma^2\sqrt{1-c_4^2}$ where c_4 is solely a function of n. Since $c_4 \neq 1$, S is a biased estimator of σ.

$$B_3 = 1 - \frac{3}{c_4}\sqrt{1-c_4^2} : \qquad B_4 = 1 + \frac{3}{c_4}\sqrt{1-c_4^2} : \qquad B_5 = c_4 - 3\sqrt{1-c_4^2} = c_4 B_3 : \qquad B_6 = c_4 + 3\sqrt{1-c_4^2} = c_4 B_4 :$$

Table 27 Tolerance Factors for the Normal Distribution

For a normal distribution with unknown mean and unknown standard deviation, a random sample of size n can be used to state with a confidence probability of 95% or 99% that the tolerance interval $\bar{x} \pm ks$ will contain a proportion of *at least* $1 - \alpha$ of the population.

$\bar{x}$ is the observed sample mean and s is the sample standard deviation $\left[= \sqrt{\dfrac{\sum (x - \bar{x})^2}{n - 1}} \right]$.

The tabulated values give the multiplier, k.

	95% Confidence Probability				99% Confidence Probability		
	Proportion $(1 - \alpha)$				Proportion $(1 - \alpha)$		
n	0.90	0.95	0.99	n	0.90	0.95	0.99
2	32.019	37.764	48.430	2	160.193	188.491	242.300
3	8.380	9.916	12.861	3	18.930	22.401	29.055
4	5.369	6.370	8.299	4	9.398	11.150	14.527
5	4.275	5.079	6.634	5	6.612	7.855	10.260
6	3.712	4.414	5.775	6	5.337	6.345	8.301
7	3.369	4.007	5.248	7	4.613	5.488	7.187
8	3.136	3.732	4.891	8	4.147	4.936	6.468
9	2.967	3.532	4.631	9	3.822	4.550	5.966
10	2.839	3.379	4.433	10	3.582	4.265	5.594
11	2.737	3.259	4.277	11	3.397	4.045	5.308
12	2.655	3.162	4.150	12	3.250	3.870	5.079
13	2.587	3.081	4.044	13	3.130	3.727	4.893
14	2.529	3.012	3.955	14	3.029	3.608	4.737
15	2.480	2.954	3.878	15	2.945	3.507	4.605
16	2.437	2.903	3.812	16	2.872	3.421	4.492
17	2.400	2.858	3.754	17	2.808	3.345	4.393
18	2.366	2.819	3.702	18	2.753	3.279	4.307
19	2.337	2.784	3.656	19	2.703	3.221	4.230
20	2.310	2.752	3.615	20	2.659	3.168	4.161
25	2.208	2.631	3.457	25	2.494	2.972	3.904
30	2.140	2.549	3.350	30	2.385	2.841	3.733
35	2.090	2.490	3.272	35	2.306	2.748	3.611
40	2.052	2.445	3.213	40	2.247	2.677	3.518
45	2.021	2.408	3.165	45	2.200	2.621	3.444
50	1.996	2.379	3.126	50	2.162	2.576	3.385
55	1.976	2.354	3.094	55	2.130	2.538	3.335
60	1.958	2.333	3.066	60	2.103	2.506	3.293
65	1.943	2.315	3.042	65	2.080	2.478	3.257
70	1.929	2.299	3.021	70	2.060	2.454	3.225
75	1.917	2.285	3.002	75	2.042	2.433	3.197
80	1.907	2.272	2.986	80	2.026	2.414	3.173
85	1.897	2.261	2.971	85	2.012	2.397	3.150
90	1.889	2.251	2.958	90	1.999	2.382	3.130
95	1.881	2.241	2.945	95	1.987	2.368	3.112
100	1.874	2.233	2.934	100	1.977	2.355	3.096
150	1.825	2.175	2.859	150	1.905	2.270	2.983
200	1.798	2.143	2.816	200	1.865	2.222	2.921
250	1.780	2.121	2.788	250	1.839	2.191	2.880
300	1.767	2.106	2.767	300	1.820	2.169	2.850
400	1.749	2.084	2.739	400	1.794	2.138	2.809
500	1.737	2.070	2.721	500	1.777	2.117	2.783
600	1.729	2.060	2.707	600	1.764	2.102	2.763
700	1.722	2.052	2.697	700	1.755	2.091	2.748
800	1.717	2.046	2.688	800	1.747	2.082	2.736
900	1.712	2.040	2.682	900	1.741	2.075	2.726
1000	1.709	2.036	2.676	1000	1.736	2.068	2.718
∞	1.645	1.960	2.576	∞	1.645	1.960	2.576

Table 28 Sample Size for Two-sided Distribution-free Tolerance Limits

The table gives the size of a random sample to be taken from *any* underlying distribution in order to ensure that there is a given confidence probability of $100(1 - \gamma)$% that at least a proportion $(1 - \alpha)$ of the members of the distribution will lie between the smallest and largest values subsequently observed in the sample.

Proportion	Confidence Probability: $100(1 - \gamma)$%					
$1 - \alpha$	50%	70%	90%	95%	99%	99.5%
0.995	336	488	777	947	1325	1483
0.99	168	244	388	473	662	740
0.95	34	49	77	93	130	146
0.90	17	24	38	46	64	72
0.85	11	16	25	30	42	47
0.80	9	12	18	22	31	34
0.75	7	10	15	18	24	27
0.70	6	8	12	14	20	22
0.60	4	6	9	10	14	16
0.50	3	5	7	8	11	12

Table 29 Sample Size for One-sided Distribution-free Tolerance Limits

The table gives the size of a random sample to be taken from any underlying distribution in order to ensure that there is a given confidence probability of $100(1 - \gamma)$% that at least a proportion $(1 - \alpha)$ of the members of the distribution will be greater than the smallest (or less than the largest) value subsequently observed in the sample.

Proportion	Confidence Probability: $100(1 - \gamma)$%				
$1 - \alpha$	50	70	95	99	99.5
0.995	139	241	598	919	1379
0.99	69	120	299	459	688
0.95	14	24	59	90	135
0.90	7	12	29	44	66
0.85	5	8	19	29	43
0.80	4	6	14	21	31
0.75	3	5	11	17	25
0.70	2	4	9	13	20
0.60	2	3	6	10	14
0.50	1	2	5	7	10

Notes on Tables 30, 31, 32, 33, 34

These tables are generally applicable when testing the randomness of a sequence of observations by means of a classification of the sequence into runs of various kinds.

A discussion of the theory of run tests of significance and their application is given in a number of textbooks on statistics and quality control. Such textbooks include:

Statistical Theory with Engineering Applications A. Hald, J. Wiley Inc.

Statistical Theory and Methodology in Science and Engineering K.A. Brownlee, J. Wiley Inc.

Quality Control and Industrial Statistics A.J. Duncan, Richard D. Irwin.

and themselves contain references to the major papers on the subject.

Example

In order to illustrate the use of tables, the thirty numbers given below might represent successive measurements of some quantity recorded in the time sequence in which they were obtained.

0.030, 0.640, 0.022, 1.133, 0.580, 0.370, 0.334, 0.455, 0.807, 1.886, 0.287, 1.413, 0.463, 0.095, 2.495, 1.902, 2.210, 0.037, 0.341, 1.044, 0.120, 1.155, 0.452, 0.160, 0.040, 0.038, 0.152, 0.433, 1.104, 0.096.

The average of the readings is 0.676 and the median falls between 0.433 and 0.452. Each observation can then be classified as being *above* or *below* the average; alternatively, it can be classified as being *above* or *below* the median.

A run is defined as a succession of points which are of the same kind, the number of such points determining the length of the run. Thus considering the example, the sequence starts with a run of length 3 points below average, a run of 1 point above average, then a run of 4 points below average and so on giving a final total of 7 runs above average and 8 runs below, 15 runs in all; there are 10 individual readings above average and 20 below.

If the underlying population average has been changing gradually during the period of time covered by the thirty measurements, then apart from local sampling fluctuations, the figures should tend to increase or decrease steadily. In this case there will tend to be few runs above and below the average and the runs will tend to be long. On the other hand, if the figures represent a random sequence of observations from a stable population, then a larger number of shorter runs would be expected. With respect to the median as reference point, there are 8 runs above the median and 9 runs below, the longest run being of length 5.

Tables 30 and 31

The 30 sample points are divided into 10 above the average and 20 below and yield a total of 15 runs. Reference to Table 30 for $m = 10$ and $n = 20$ shows that the probability of obtaining 9 or fewer runs with such a split of points is at most 5% when the hypothesis of random grouping is assumed. Since the observed value of 15 runs is greater than 9, the hypothesis cannot be rejected at the 5% level. If the observed number of runs had been less than 9, Table 31 would show whether or not this number was significant at the 0.5% level. Note that, because of the discrete nature of the distribution of the number of runs, the actual probabilities of an equal or smaller number of runs than tabulated do not exceed the quoted values; they will usually be smaller than 5% and 0.5% respectively.

Tables 32 and 33

With respect to the median as reference point, there are 8 runs above and 9 runs below the median, the longest run being of length 5. In this case $n = 15$ and Table 32 shows that the observed total of 17 runs is not small enough to reject the hypothesis of randomness of grouping at either the 5% (11 runs) or the 0.5% (8 runs) levels.

Table 33 shows that for 30 points, there is a 5% chance that the longest run above or below the median will be 8 or more and a 1% chance that it will be 9 or more. Since the longest observed run is 5, the hypothesis of randomness of grouping is not discredited. Note that the two tests using Tables 32 and 33 are not independent.

Table 34

In cases where the underlying population mean may have been moving in a cyclic fashion during the sampling period, it is convenient to consider the number and lengths of runs up and down. A sequence of continually increasing values lead to a run up while a run down is given by a sequence of continually decreasing values. The 30 points will give 29 successive differences which will be either positive or negative (zero differences are usually ignored). Reference to the original data shows that the sequence of differences begins $+ - + - - - + + + - \ldots$ and that the longest run is a run down of length 4. Table 34 shows that for $N = 30$, the probability of a run up or down of length 7 or more is about 0.1% while a run of length 5 or more will occur with probability of about 5%. Again the hypothesis of randomness is not rejected.

Note that too many very short runs would also be indicative of non-randomness but the tables do not allow evaluation of this situation.

Table 30 Number of Runs on Either Side of the Mean, 5.0% Point

Table for Testing Randomness of Grouping in a Sequence of Alternatives

(Probability of an equal or smaller number of runs than that listed is $P = 0.05$)
n = cases on one side of average: m = cases on other side of average
m is always taken as the smaller number of cases, n the larger.

$m =$	6	7	8	9	10	11	12	13	14	15	16	17	18	19	20
$n = 6$	3														
7	4	4													
8	4	4	5												
9	4	5	5	6											
10	5	5	6	6	6										
11	5	5	6	6	7	7									
12	5	6	6	7	7	8	8								
13	5	6	6	7	8	8	9	9							
14	5	6	7	7	8	8	9	9	10						
15	6	6	7	8	8	9	9	10	10	11					
16	6	6	7	8	8	9	10	10	11	11	11				
17	6	7	7	8	9	9	10	10	11	11	12	12			
18	6	7	8	8	9	10	10	11	11	12	12	13	13		
19	6	7	8	8	9	10	10	11	12	12	13	13	14	14	
20	6	7	8	9	9	10	11	11	12	12	13	13	14	14	15

Table 31 Number of Runs on Either Side of the Mean, 0.5% Point

Table for Testing Randomness of Grouping in a Sequence of Alternatives

(Probability of an equal or smaller number of runs than that listed is $P = 0.005$)
n = cases on one side of average: m = cases on other side of average
m is always taken as the smaller number of cases, n the larger.

$m =$	6	7	8	9	10	11	12	13	14	15	16	17	18	19	20
$n = 6$	2														
7	2	3													
8	3	3	3												
9	3	3	3	4											
10	3	3	4	4	5										
11	3	4	4	5	5	5									
12	3	4	4	5	5	6	6								
13	3	4	5	5	5	6	6	7							
14	4	4	5	5	6	6	7	7	7						
15	4	4	5	6	6	7	7	7	8	8					
16	4	5	5	6	6	7	7	8	8	9	9				
17	4	5	5	6	7	7	8	8	8	9	9	10			
18	4	5	6	6	7	7	8	8	9	9	10	10	11		
19	4	5	6	6	7	8	8	9	9	10	10	10	11	11	
20	4	5	6	7	7	8	8	9	9	10	10	11	11	12	12

Both tables after Freda S. Swed and C. Eisenhart, 'Tables for Testing Randomness of Grouping in a Sequence of Alternatives', *Annals of Mathematical Statistics*, Vol. XIV (1943), pp 66-68. Reprinted with permission of the Institute of Mathematical Statistics.

Table 32 Number of Runs Above and Below the Median

Limiting values for the total number of runs above and below the median of a set of values for $P = 0.005$ and $P = 0.05$ where P is the probability of an equal or smaller total number of runs.

$m = n$	0.005	0.05	$m = n$	0.005	0.05
10	4	6	55	42	46
11	5	7	56	42	47
12	6	8	57	43	48
13	7	9	58	44	49
14	7	10	59	45	50
15	8	11			
16	9	11	60	46	51
17	10	12	61	47	52
18	10	13	62	48	53
19	11	14	63	49	54
			64	49	55
20	12	15	65	50	56
21	13	16	66	51	57
22	14	17	67	52	58
23	14	17	68	53	58
24	15	16	69	54	59
25	16	19			
26	17	20	70	55	60
27	18	21	71	56	61
28	18	22	72	57	62
29	19	23	73	57	63
			74	58	64
30	20	24	75	59	65
31	21	25	76	60	66
32	22	25	77	61	67
33	23	26	78	62	68
34	23	27	79	63	69
35	24	28			
36	25	29	80	64	70
37	26	30	81	65	71
38	27	31	82	66	71
39	28	32	83	66	72
			84	67	73
40	29	33	85	68	74
41	29	34	86	69	75
42	30	35	87	70	76
43	31	35	88	71	77
44	32	36	89	72	78
45	33	37			
46	34	38	90	73	79
47	35	39	91	74	80
48	35	40	92	75	81
49	36	41	93	75	82
			94	76	83
50	37	42	95	77	84
51	38	43	96	78	85
52	39	44	97	79	86
53	40	45	98	80	87
54	41	45	99	81	87
			100	82	88

Table after Freda S. Swed and C. Eisenhart, 'Tables for Testing Randomness of Grouping in a Sequence of Alternatives', *Annals of Mathematical Statistics*, Vol. XIV (1943), pp 66-68. Reprinted with permission of the Institute of Mathematical Statistics.

Table 33 Lengths of Runs on Either Side of the Median

Probability of getting at least one run of specified size or more.

N	0.05	0.01	0.001
10	5	—	—
20	7	8	9
30	8	9	—
40	9	10	12
50	10	11	—

(Larger runs than these suggest the existence of non-random influences.)

Table after F. Mosteller, 'Note on Application of Runs to Quality Control Charts', *Annals of Mathematical Statistics*, Vol. XII (1941), p 232. Reprinted with permission of the Institute of Mathematical Statistics.

Table 34 Critical Values of Lengths of Runs Up and Down

N	Probability equal to or less than 0.0032		Probability equal to or less than 0.0583	
	Run	Probability of an equal or greater run	Run	Probability of an equal or greater run
4	–	–	3	0.0833
5	–	–	4	0.0167
6	5	0.0028	4	0.0306
7	6	0.0004	4	0.0444
8	6	0.0007	4	0.0583
9	6	0.0011	5	0.0099
10	6	0.0014	5	0.0123
11	6	0.0018	5	0.0147
12	6	0.0021	5	0.0170
13	6	0.0025	5	0.0194
14	6	0.0028	5	0.0217
15	6	0.0032*	5	0.0239*
20	7	0.0006*	5	0.0355*
40	7	0.0015*	6	0.0118*
60	7	0.0023*	6	0.0186*
80	7	0.0032*	6	0.0254*
100	8	0.0005*	6	0.0322*
200	8	0.0010*	7	0.0085*
500	8	0.0024*	7	0.0215*
1,000	9	0.0005*	7	0.0428*
5,000	9	0.0025*	8	0.0245*

*Probabilities based on approximation of the exact distribution by the Poisson exponential. See P. S. Olmstead, 'Distribution of Sample Arrangements for Runs Up and Down', *Annals of Mathematical Statistics*, Vol. XVII (1946), p29. Reprinted with permission of the Institute of Mathematical Statistics.

Table 35 Derivation of Single Sampling Plans

Values of np_1 and c for constructing single sampling plans whose OC Curve is required to pass through the two points $(p_1, 1 - \alpha)$ and (p_2, β).

Here p_1 is the fraction non-conforming for which the risk of wrongful rejection is to be α, and p_2 is the fraction non-conforming for which the risk of wrongful acceptance is to be β ($p_2 > p_1$). To construct the plan, find the tabular value of p_2/p_1 in the column for the given α and β which is equal to or just greater than the given value of the ratio. The sample size is found by dividing the value of np_1 corresponding to the selected ratio by p_1. The acceptance number is the value of c corresponding to the selected value of the ratio.

| | Values of p_2/p_1 for | | | | | Values of p_2/p_1 for | | | |
| | $\alpha = 0.05$ | $\alpha = 0.05$ | $\alpha = 0.05$ | | | $\alpha = 0.01$ | $\alpha = 0.01$ | $\alpha = 0.01$ | |
c	$\beta = 0.10$	$\beta = 0.05$	$\beta = 0.01$	np_1	c	$\beta = 0.10$	$\beta = 0.05$	$\beta = 0.01$	np_1
0	44.890	58.404	89.781	0.052	0	229.105	298.073	458.210	0.010
1	10.946	13.349	18.681	0.355	1	26.184	31.933	44.686	0.149
2	6.509	7.699	10.280	0.818	2	12.206	14.439	19.278	0.436
3	4.890	5.675	7.352	1.366	3	8.115	9.418	12.202	0.823
4	4.057	4.646	5.890	1.970	4	6.249	7.156	9.072	1.279
5	3.549	4.023	5.017	2.613	5	5.195	5.889	7.343	1.785
6	3.206	3.604	4.435	3.286	6	4.520	5.082	6.253	2.330
7	2.957	3.303	4.019	3.981	7	4.050	4.524	5.506	2.906
8	2.768	3.074	3.707	4.695	8	3.705	4.115	4.962	3.507
9	2.618	2.895	3.462	5.426	9	3.440	3.803	4.548	4.130
10	2.497	2.750	3.265	6.169	10	3.229	3.555	4.222	4.771
11	2.397	2.630	3.104	6.924	11	3.058	3.354	3.959	5.428
12	2.312	2.528	2.968	7.690	12	2.915	3.188	3.742	6.099
13	2.240	2.442	2.852	8.464	13	2.795	3.047	3.559	6.782
14	2.177	2.367	2.752	9.246	14	2.692	2.927	3.403	7.477
15	2.122	2.302	2.665	10.035	15	2.603	2.823	3.269	8.181
16	2.073	2.244	2.588	10.831	16	2.524	2.732	3.151	8.895
17	2.029	2.192	2.520	11.633	17	2.455	2.652	3.048	9.616
18	1.990	2.145	2.458	12.442	18	2.393	2.580	2.956	10.346
19	1.954	2.103	2.403	13.254	19.	2.337	2.516	2.874	11.082
20	1.922	2.065	2.352	14.072	20	2.287	2.458	2.799	11.825
21	1.892	2.030	2.307	14.894	21	2.241	2.405	2.733	12.574
22	1.865	1.999	2.265	15.719	22.	2.200	2.357	2.671	13.329
23	1.840	1.969	2.223	16.548	23	2.162	2.313	2.615	14.088
24	1.817	1.942	2.191	17.382	24	2.126	2.272	2.564	14.853
25	1.795	1.917	2.158	18.218	25	2.094	2.235	2.516	15.623
26	1.775	1.893	2.127	19.058	26	2.064	2.200	2.472	16.397
27	1.757	1.871	2.098	19.900	27	2.035	2.168	2.431	17.175
28	1.739	1.850	2.071	20.746	28	2.009	2.138	2.393	17.957
29	1.723	1.831	2.046	21.594	29	1.985	2.110	2.358	18.742
30	1.707	1.813	2.023	22.444	30	1.962	2.083	2.324	19.532
31	1.692	1.796	2.001	23.298	31	1.940	2.059	2.293	20.324
32	1.679	1.780	1.980	24.152	32	1.920	2.035	2.264	21.120
33	1.665	1.764	1.960	25.010	33	1.900	2.013	2.236	21.919
34	1.653	1.750	1.941	25.870	34	1.882	1.992	2.210	22.721
35	1.641	1.736	1.923	26.731	35	1.865	1.973	2.185	23.525
36	1.630	1.723	1.906	27.594	36	1.848	1.954	2.162	24.333
37	1.619	1.710	1.890	28.460	37	1.833	1.936	2.139	25.143
38	1.609	1.698	1.875	29.327	38	1.818	1.920	2.118	25.955
39	1.599	1.687	1.860	30.196	39	1.804	1.903	2.098	26.770
40	1.590	1.676	1.846	31.066	40	1.790	1.887	2.079	27.587
41	1.581	1.666	1.833	31.938	41	1.777	1.873	2.060	28.406
42	1.572	1.656	1.820	32.812	42	1.765	1.859	2.043	29.228
43	1.564	1.646	1.807	33.686	43	1.753	1.845	2.026	30.051
44	1.556	1.637	1.796	34.563	44	1.742	1.832	2.010	30.877
45	1.548	1.628	1.784	35.441	45	1.731	1.820	1.994	31.704
46	1.541	1.619	1.773	36.320	46	1.720	1.808	1.980	32.534
47	1.534	1.611	1.763	37.200	47	1.710	1.796	1.965	33.365
48	1.527	1.603	1.752	38.082	48	1.701	1.785	1.952	34.198
49	1.521	1.596	1.743	38.965	49	1.691	1.775	1.938	35.032

Table taken from J M Cameron, 'Tables for Constructing and for Computing the Operating Characteristics of Single-Sampling Plans', *Industrial Quality Control*, July 1952, pp 37-39. © 1952 American Society for Quality. Reprinted with permission.

Table 36 Construction of O.C. Curves for Single Sampling Plans

Values of np_1 for which the probability of acceptance of c or fewer non-conforming items in a sample of n is $P(A)$.

To find the non-conforming fraction p, corresponding to a probability of acceptance $P(A)$ in a single sampling plan with sample size n and acceptance number c, divide by n the entry in the row for the given c and the column for the given $P(A)$. Plot $P(A)$ against p.

$P(A)=$	0.995	0.990	0.975	0.950	0.900	0.750	0.500	0.250	0.100	0.050	0.025	0.010	0.005
$c = 0$	0.0050	0.010	0.025	0.051	0.105	0.288	0.693	1.386	2.303	2.996	3.689	4.605	5.298
1	0.103	0.149	0.242	0.355	0.532	0.961	1.678	2.693	3.890	4.744	5.572	6.638	7.430
2	0.338	0.436	0.619	0.818	1.102	1.727	2.674	3.920	5.322	6.296	7.224	8.406	9.274
3	0.672	0.823	1.090	1.366	1.745	2.535	3.672	5.109	6.681	7.754	8.768	10.045	10.978
4	1.078	1.279	1.623	1.970	2.433	3.369	4.671	6.274	7.994	9.154	10.242	11.605	12.594
5	1.537	1.785	2.202	2.613	3.152	4.219	5.670	7.423	9.275	10.513	11.668	13.108	14.150
6	2.037	2.330	2.814	3.286	3.895	5.083	6.670	8.558	10.532	11.842	13.060	14.571	15.660
7	2.571	2.906	3.454	3.981	4.656	5.956	7.669	9.684	11.771	13.148	14.422	16.000	17.134
8	3.132	3.507	4.115	4.695	5.432	6.838	8.669	10.802	12.995	14.434	15.763	17.403	18.578
9	3.717	4.130	4.795	5.426	6.221	7.726	9.669	11.914	14.206	15.705	17.085	18.783	19.998
10	4.321	4.771	5.491	6.169	7.021	8.620	10.668	13.020	15.407	16.962	18.390	20.145	21.398
11	4.943	5.428	6.201	6.924	7.829	9.519	11.668	14.121	16.598	18.208	19.682	21.490	22.779
12	5.580	6.099	6.922	7.690	8.646	10.422	12.668	15.217	17.782	19.442	20.962	22.821	24.145
13	6.231	6.782	7.654	8.464	9.470	11.329	13.668	16.310	18.958	20.668	22.230	24.139	25.496
14	6.893	7.477	8.396	9.246	10.300	12.239	14.668	17.400	20.128	21.886	23.490	25.446	26.836
15	7.566	8.181	9.144	10.035	11.135	13.152	15.668	18.486	21.292	23.098	24.741	27.743	28.166
16	8.249	8.895	9.902	10.831	11.976	14.068	16.668	19.570	22.452	24.302	25.984	28.031	29.484
17	8.942	9.616	10.666	11.633	12.822	14.986	17.668	20.652	23.606	25.500	27.220	29.310	30.792
18	9.644	10.346	11.438	12.442	13.672	15.907	18.668	21.731	24.756	26.692	28.448	30.581	32.092
19	10.353	11.082	12.216	13.254	14.525	16.830	19.668	22.808	25.902	27.879	29.671	31.845	33.383
20	11.069	11.825	12.999	14.072	15.383	17.755	20.668	23.883	27.045	29.062	30.888	33.103	34.668
21	11.791	12.574	13.787	14.894	16.244	18.682	21.668	24.956	28.184	30.241	32.102	34.355	35.947
22	12.520	13.329	14.580	15.719	17.108	19.610	22.668	26.028	29.320	31.416	33.309	35.601	37.219
23	13.255	14.088	15.377	16.548	17.975	20.540	23.668	27.098	30.453	32.586	34.512	36.841	38.485
24	13.995	14.853	16.178	17.382	18.844	21.471	24.668	28.167	31.584	33.752	35.710	38.077	39.745
25	14.740	15.623	16.984	18.218	19.717	22.404	25.667	29.234	32.711	34.916	36.905	39.308	41.000
26	15.490	16.397	17.793	19.058	20.592	23.338	26.667	30.300	33.836	36.077	38.096	40.535	42.252
27	16.245	17.175	18.606	19.900	21.469	24.273	27.667	31.365	34.959	37.234	39.284	41.757	43.497
28	17.004	17.957	19.422	20.746	22.348	25.209	28.667	32.428	36.080	38.389	40.468	42.975	44.738
29	17.767	18.742	20.241	21.594	23.229	26.147	29.667	33.491	37.198	39.541	41.649	44.190	45.976
30	18.534	19.532	21.063	22.444	24.113	27.086	30.667	34.552	38.315	40.690	42.827	45.401	47.210
31	19.305	20.324	21.888	23.298	24.998	28.025	31.667	35.613	39.430	41.838	44.002	46.609	48.440
32	20.079	21.120	22.716	24.152	25.885	28.966	32.667	36.672	40.543	42.982	45.174	47.813	49.666
33	20.856	21.919	23.546	25.010	26.774	29.907	33.667	37.731	41.654	44.125	46.344	49.015	50.888
34	21.638	22.721	24.379	25.870	27.664	30.849	34.667	38.788	42.764	45.266	47.512	50.213	52.108
35	22.422	23.525	25.214	26.731	28.556	31.792	35.667	39.845	43.872	46.404	48.676	51.409	53.324
36	23.208	24.333	26.052	27.594	29.450	32.736	36.667	40.901	44.978	47.540	49.840	52.601	54.538
37	23.998	25.143	26.891	28.460	30.345	33.681	37.667	41.957	46.083	48.676	51.000	53.791	55.748
38	24.791	25.955	27.733	29.327	31.241	34.626	38.667	43.011	47.187	49.808	52.158	54.979	56.956
39	25.586	26.770	28.576	30.196	32.139	35.572	39.667	44.065	48.289	50.940	53.314	56.164	58.160
40	26.384	27.587	29.422	31.066	33.038	36.519	40.667	45.118	49.390	52.069	54.469	57.347	59.363
41	27.184	28.406	30.270	31.938	33.938	37.466	41.667	46.171	50.490	53.197	55.622	58.528	60.563
42	27.986	29.228	31.120	32.812	34.839	38.414	42.667	47.223	51.589	54.324	56.772	59.717	61.761
43	28.791	30.051	31.970	33.686	35.742	39.363	43.667	48.274	52.686	55.449	57.921	60.884	62.956
44	29.598	30.877	32.824	34.563	36.646	40.312	44.667	49.325	53.782	56.572	59.068	62.059	64.150
45	30.408	31.704	33.678	35.441	37.550	41.262	45.667	50.375	54.878	57.695	60.214	63.231	65.340
46	31.219	32.534	34.534	36.320	38.456	42.212	46.667	51.425	55.972	58.816	61.358	64.402	66.529
47	32.032	33.365	35.392	37.200	39.363	43.163	47.667	52.474	57.064	59.936	62.500	65.571	67.716
48	32.848	34.198	36.250	38.082	40.270	44.115	48.667	53.522	58.158	61.054	63.641	66.738	68.901
49	33.664	35.032	37.111	38.965	41.179	45.067	49.667	54.571	59.249	62.171	64.780	67.903	70.084

Table taken from J M Cameron, 'Tables for Constructing and for Computing the Operating Characteristic of Single-Sampling Plans', *Industrial Quality Control*, July 1952, pp 37-39. © 1952 American Society for Quality. Reprinted with permission.

Table 37 Random Numbers (1)

03 47 43 73 86	36 96 47 36 61	46 98 63 71 62	33 26 16 80 45	60 11 14 10 95
97 74 24 67 62	42 81 14 57 20	42 53 32 37 32	27 07 36 07 51	24 51 79 89 73
16 76 62 27 66	56 50 26 71 07	32 90 79 78 53	13 55 38 58 59	88 97 54 14 10
12 56 85 99 26	96 96 68 27 31	05 03 72 93 15	57 12 10 14 21	88 26 49 81 76
55 59 56 35 64	38 54 82 46 22	31 62 43 09 90	06 18 44 32 53	23 83 01 30 30
16 22 77 94 39	49 54 43 54 82	17 37 93 23 78	87 35 20 96 43	84 26 34 91 64
84 42 17 53 31	57 24 55 06 88	77 04 74 47 67	21 76 33 50 25	83 92 12 06 76
63 01 63 78 59	16 95 55 67 19	98 10 50 71 75	12 86 73 58 07	44 39 52 38 79
33 21 12 34 29	78 64 56 07 82	52 42 07 44 38	15 51 00 13 42	99 66 02 79 54
57 60 86 32 44	09 47 27 96 54	49 17 46 09 62	90 52 84 77 27	08 02 73 43 28
18 18 07 92 46	44 17 16 58 09	79 83 86 19 62	06 76 50 03 10	55 23 64 05 05
26 62 38 97 75	84 16 07 44 99	83 11 46 32 24	20 14 85 88 45	10 93 72 88 71
23 42 40 64 74	82 97 77 77 81	07 45 32 14 08	32 98 94 07 72	93 85 79 10 75
52 36 28 19 95	50 92 26 11 97	00 56 76 31 38	80 22 02 53 53	86 60 42 04 53
37 85 94 35 12	83 39 50 08 30	42 34 07 96 88	54 42 06 87 98	35 85 29 48 39
70 29 17 12 13	40 33 20 38 26	13 89 51 03 74	17 76 37 13 04	07 74 21 19 30
56 62 18 37 35	96 83 50 87 75	97 12 25 93 47	70 33 24 03 54	97 77 46 44 80
99 49 57 22 77	88 42 95 45 72	16 64 36 16 00	04 43 18 66 79	94 77 24 21 90
16 08 15 04 72	33 27 14 34 09	45 59 34 68 49	12 72 07 34 45	99 27 72 95 14
31 16 93 32 43	50 27 89 87 19	20 15 37 00 49	52 85 66 60 44	38 68 88 11 80
68 34 30 13 70	55 74 30 77 40	44 22 78 84 26	04 33 46 09 52	68 07 97 06 57
74 57 25 65 76	59 29 97 68 60	71 91 38 67 54	13 58 18 24 76	15 54 55 95 52
27 42 37 86 53	48 55 90 65 72	96 57 69 36 10	96 46 92 42 45	97 60 49 04 91
00 39 68 29 61	66 37 32 20 30	77 84 57 03 29	10 45 65 04 26	11 04 96 67 24
29 94 98 94 24	68 49 69 10 82	53 75 91 93 30	34 25 20 57 27	40 48 73 51 92
16 90 82 66 59	83 62 64 11 12	67 19 00 72 74	60 57 21 29 68	02 02 37 03 31
11 27 94 75 06	06 09 19 74 66	02 94 37 34 02	76 70 90 30 86	38 45 94 30 38
35 24 10 16 20	33 32 51 26 38	79 78 45 04 91	16 92 53 56 16	02 75 50 95 98
38 23 16 86 38	42 38 97 01 50	87 75 66 81 41	40 01 74 91 62	48 51 84 08 32
31 96 25 91 47	96 44 33 49 13	34 86 82 53 91	00 52 43 48 85	27 55 26 89 62
66 67 40 67 14	64 05 71 95 86	11 05 65 09 68	76 83 20 37 90	57 16 00 11 66
14 90 84 45 11	75 73 88 05 90	52 27 41 14 86	22 98 12 22 08	07 52 74 95 80
68 05 51 18 00	33 96 02 75 19	07 60 62 93 55	59 33 82 43 90	49 37 38 44 59
20 46 78 73 90	97 51 40 14 02	04 02 33 31 08	39 54 16 49 36	47 95 93 13 30
64 19 58 97 79	15 06 15 93 20	01 90 10 75 06	40 78 78 89 62	02 67 74 17 33
05 26 93 70 60	22 35 85 15 13	92 03 51 59 77	59 56 78 06 83	52 91 05 70 74
07 97 10 88 23	09 98 42 99 64	61 71 62 99 15	06 51 29 16 93	58 05 77 08 51
68 71 86 85 85	54 87 66 47 54	73 32 08 11 12	44 95 92 63 16	29 56 24 29 48
26 99 61 65 53	58 37 78 80 70	42 10 50 67 53	32 17 55 85 74	94 44 67 16 94
14 65 52 68 75	87 59 36 22 41	26 78 63 06 55	13 08 27 01 50	15 29 39 39 43
17 53 77 58 71	71 41 61 50 72	12 41 94 96 26	44 95 27 36 99	02 96 74 30 83
90 26 59 21 19	23 52 23 33 12	95 93 02 18 39	07 02 18 36 07	25 99 32 70 23
41 23 52 55 99	31 04 49 69 96	10 47 48 45 88	13 41 43 89 20	97 17 14 49 17
60 20 50 81 69	31 99 73 68 68	35 81 33 03 76	24 30 12 48 60	18 99 10 72 34
91 25 38 05 90	94 58 28 41 36	45 37 59 03 09	90 35 57 29 12	82 62 54 65 60
34 50 57 74 37	98 80 33 00 91	09 77 93 19 82	74 94 80 04 04	45 07 31 66 49
85 22 04 39 43	73 81 53 94 79	33 62 46 86 28	08 31 54 46 31	53 94 13 38 47
09 79 13 77 48	73 82 97 22 21	05 03 27 24 83	72 89 44 05 60	35 80 39 94 88
88 75 80 18 14	22 95 75 42 49	39 32 82 22 49	02 48 07 70 37	16 04 61 67 87
90 96 23 70 00	39 00 03 06 90	55 85 78 38 36	94 37 30 69 32	90 89 00 76 33

This table is taken from Table XXXIII of Fisher and Yates: *Statistical Tables for Biological Agricultural and Medical Research*, reprinted by permission of Addison Wesley Longman Ltd.

Table 37 Random Numbers (2) – continued

53 74 23 99 67	61 32 28 69 84	94 62 67 86 24	98 33 41 19 95	47 53 53 38 09
63 38 06 86 54	99 00 65 26 94	02 82 90 23 07	79 62 67 80 60	75 91 12 81 19
35 30 58 21 46	06 72 17 10 94	25 21 31 75 96	49 28 24 00 49	55 65 79 78 07
63 43 36 82 69	65 51 18 37 88	61 38 44 12 45	32 92 85 88 65	54 34 81 85 35
98 25 37 55 26	01 91 82 81 46	74 71 12 94 97	24 02 71 37 07	03 92 18 66 75
02 63 21 17 69	71 50 80 89 56	38 15 70 11 48	43 40 45 86 98	00 83 26 91 03
64 55 22 21 82	48 22 28 06 00	61 54 13 43 91	82 78 12 23 29	06 66 24 12 27
85 07 26 13 89	01 10 07 82 04	59 63 69 36 03	69 11 15 83 80	13 29 54 19 28
58 54 16 24 15	51 54 44 82 00	62 61 65 04 69	38 18 65 18 97	85 72 13 49 21
34 85 27 84 87	61 48 64 56 26	90 18 48 13 26	37 70 15 42 57	65 65 80 39 07
03 92 18 27 46	57 99 16 96 56	30 33 72 85 22	84 64 38 56 98	99 01 30 98 64
62 95 30 27 59	37 75 41 66 48	86 97 80 61 45	23 53 04 01 63	45 76 08 64 27
08 45 93 15 22	60 21 75 46 91	98 77 27 85 42	28 88 61 08 84	69 62 03 42 73
07 08 55 18 40	45 44 75 13 90	24 94 96 61 02	57 55 66 83 15	73 42 37 11 61
01 85 89 95 66	51 10 19 34 88	15 84 97 19 75	12 76 39 43 78	64 63 91 08 25
72 84 71 14 35	19 11 58 49 26	50 11 17 17 76	86 31 57 20 18	95 60 78 46 75
88 78 28 16 84	13 52 53 94 53	75 45 69 30 96	73 89 65 70 31	99 17 43 48 76
45 17 75 65 57	28 40 19 72 12	25 12 74 75 67	60 40 60 81 19	24 62 01 61 16
96 76 28 12 54	22 01 11 94 25	71 96 16 16 88	68 64 36 74 45	19 59 50 88 92
43 31 67 72 30	24 02 94 08 63	38 32 36 66 02	69 36 38 25 39	48 03 45 15 22
50 44 66 44 21	66 06 58 05 62	68 15 54 35 02	42 35 48 96 32	14 52 41 52 48
22 66 22 15 86	26 63 75 41 99	58 42 36 72 24	58 37 52 18 51	03 37 18 39 11
96 24 40 14 51	23 22 30 88 57	95 67 47 29 83	94 69 40 06 07	18 16 36 78 86
31 73 91 61 19	60 20 72 98 48	98 57 07 23 69	65 95 39 69 58	56 80 30 19 44
78 60 73 99 84	43 89 94 36 45	56 69 47 07 41	90 22 91 07 12	78 35 34 08 72
84 37 90 61 56	70 10 23 98 05	85 11 34 76 60	76 48 45 34 60	01 64 18 39 96
36 67 10 08 23	98 93 35 08 86	99 29 76 29 81	33 34 91 58 93	63 14 52 32 52
07 28 59 07 48	89 64 58 89 75	83 85 62 27 89	30 14 78 56 27	86 63 59 80 02
10 15 83 87 60	79 24 31 66 56	21 48 24 06 93	91 98 94 05 49	01 47 59 38 00
55 19 68 97 65	03 73 52 16 56	00 53 55 90 27	33 42 29 38 87	22 13 88 83 34
53 81 29 13 39	35 01 20 71 34	62 33 74 82 14	53 73 19 09 03	56 54 29 56 93
51 86 32 68 92	33 98 74 66 99	40 14 71 94 58	45 94 19 38 81	14 44 99 81 07
35 91 70 29 13	80 03 54 07 27	96 94 78 32 66	50 95 52 74 33	13 80 55 62 54
37 71 67 95 13	20 02 44 95 94	64 85 04 05 72	01 32 90 76 14	53 89 74 60 41
93 66 13 83 27	92 79 64 64 72	28 54 96 53 84	48 14 52 98 94	56 07 93 89 30
02 96 08 45 65	13 05 00 41 84	93 07 54 72 59	21 45 57 09 77	19 48 56 27 44
49 83 43 48 35	82 88 33 69 96	72 36 04 19 76	4745 15 18 60	82 11 08 95 97
84 60 71 62 46	40 80 81 30 37	34 39 23 05 38	25 15 35 71 30	88 12 57 21 77
18 71 30 88 71	44 91 14 88 47	89 23 30 63 15	56 34 20 47 89	99 82 93 24 98
79 69 10 61 78	71 32 76 95 62	87 00 22 58 40	92 54 01 75 25	43 11 71 99 31
75 93 36 57 83	56 20 14 82 11	74 21 97 90 65	96 42 68 63 86	74 54 13 26 94
38 30 92 29 03	06 28 81 39 38	62 25 06 84 63	61 29 08 93 67	04 32 92 08 09
51 29 50 10 34	31 57 75 95 80	51 97 02 74 77	76 15 48 49 44	18 55 63 77 09
21 31 38 86 24	37 79 81 53 74	73 24 16 10 33	52 83 90 94 76	70 47 14 54 36
29 01 23 87 88	58 02 39 37 67	42 10 14 20 92	16 55 23 42 45	54 96 09 11 06
95 33 95 22 00	18 74 72 00 18	38 79 58 69 32	81 76 80 26 92	82 80 84 25 39
90 84 60 79 80	24 36 59 87 38	82 07 53 89 35	96 35 23 79 18	05 98 90 07 35
46 40 62 98 82	54 97 20 56 95	15 74 80 08 32	16 46 70 50 80	67 72 16 42 79
20 31 89 03 43	38 46 82 68 72	32 14 82 99 70	80 60 47 18 97	63 49 30 21 30
71 59 73 05 50	08 22 23 71 77	91 01 93 20 49	82 96 59 26 94	66 39 67 98 60

Table 37 Random Numbers (3) – continued

22 17 66 65 84	68 95 23 92 35	87 02 22 57 51	61 09 43 95 06	58 24 82 03 47
19 36 27 59 46	13 79 93 37 55	39 77 32 77 09	85 52 05 30 62	47 83 51 62 74
16 77 23 02 77	09 61 87 25 21	28 06 24 25 93	16 71 13 59 78	23 05 47 47 25
78 43 76 71 61	20 44 90 32 64	97 67 63 99 61	46 38 03 93 22	69 81 21 99 21
03 28 28 26 08	73 37 32 04 05	69 30 16 09 05	88 69 58 28 99	35 07 44 75 47
93 22 53 64 39	07 10 63 76 35	87 03 04 79 88	08 13 13 85 51	55 34 57 72 69
78 76 58 54 74	92 38 70 96 92	52 06 79 79 45	82 63 18 27 44	69 66 92 19 09
23 68 35 26 00	99 53 93 61 28	52 70 05 48 34	56 65 05 61 86	90 92 10 70 80
15 39 25 70 99	93 86 52 77 65	15 33 59 05 28	22 87 26 07 47	86 96 98 29 06
58 71 96 30 24	18 46 23 34 27	85 13 99 24 44	49 18 09 79 49	74 16 32 23 02
57 35 27 33 72	24 53 63 94 09	41 10 76 47 91	44 04 95 49 66	39 60 04 59 81
48 50 86 54 48	22 06 34 72 52	82 21 15 65 20	33 29 94 71 11	15 91 29 12 03
61 96 48 95 03	07 16 39 33 66	98 56 10 56 79	77 21 30 27 12	90 49 22 23 62
36 93 89 41 26	29 70 83 63 51	99 74 20 52 36	87 09 41 15 09	98 60 16 03 03
18 87 00 42 31	57 90 12 02 07	23 47 37 17 31	54 08 01 88 63	39 41 88 92 10
88 56 53 27 59	33 35 72 67 47	77 34 55 45 70	08 18 27 38 90	16 95 86 70 75
09 72 95 84 29	49 41 31 06 70	42 38 06 45 18	64 84 73 31 65	52 53 37 97 15
12 96 88 17 31	65 19 69 02 83	60 75 86 90 68	24 64 19 35 51	56 61 87 39 12
85 94 57 24 16	92 09 84 38 76	22 00 27 69 85	29 81 94 78 70	21 94 47 90 12
38 64 43 59 98	98 77 87 68 07	91 51 67 62 44	40 98 05 93 78	23 32 65 41 18
53 44 09 42 72	00 41 85 79 79	68 47 22 00 20	35 55 31 51 51	00 83 63 22 55
40 76 66 26 84	57 99 99 90 37	36 63 32 08 58	37 40 13 68 97	87 64 81 07 83
02 17 79 18 05	12 59 52 57 02	22 07 90 47 03	28 14 11 30 79	20 69 22 40 98
95 17 82 06 53	31 51 10 96 46	92 06 88 07 77	56 11 50 81 69	40 23 72 51 39
35 76 22 42 92	96 11 83 44 80	34 68 35 48 77	33 42 40 90 60	73 96 53 97 86
26 29 13 56 41	85 47 04 66 08	34 72 57 59 13	82 43 80 46 15	38 26 61 70 04
77 80 20 75 82	72 82 32 99 90	63 95 73 76 63	89 73 44 99 05	48 67 26 43 18
46 40 66 44 52	91 36 74 43 53	30 82 13 54 00	78 45 63 98 35	55 03 36 67 68
37 56 08 18 09	77 53 84 46 47	31 91 18 95 58	24 16 74 11 53	44 10 13 85 57
61 65 61 68 66	37 27 47 39 19	84 83 70 07 48	53 21 40 06 71	95 06 79 88 54
93 43 69 64 07	34 18 04 52 35	56 27 09 24 86	61 85 53 83 45	19 90 70 99 00
21 96 60 12 99	11 20 99 45 18	48 13 93 55 34	18 37 79 49 90	65 97 38 20 46
95 20 47 97 97	27 37 83 28 71	00 06 41 41 74	45 89 09 39 84	51 67 11 52 49
97 86 21 78 73	10 65 81 92 59	58 76 17 14 97	04 76 62 16 17	17 95 70 45 80
69 92 06 34 13	59 71 74 17 32	27 55 10 24 19	23 71 82 13 74	63 52 52 01 41
04 31 17 21 56	33 73 99 19 87	26 72 39 27 67	53 77 57 68 93	60 61 97 22 61
61 06 98 03 91	87 14 77 43 96	43 00 65 98 50	45 60 33 01 07	98 99 46 50 47
85 93 85 86 88	72 87 08 62 40	16 06 10 89 20	23 21 34 74 97	76 38 03 29 63
21 74 32 47 45	73 96 07 94 52	09 65 90 77 47	25 76 16 19 33	53 05 70 53 30
15 69 53 82 80	79 96 23 53 10	65 39 07 16 29	45 33 02 43 70	02 87 40 41 45
02 89 08 04 49	20 21 14 68 86	87 63 93 95 17	11 29 01 95 80	35 14 97 35 33
87 18 15 89 79	85 43 01 72 73	08 61 74 51 69	89 74 39 82 15	94 51 33 41 67
98 83 71 94 22	59 97 50 99 52	08 52 85 08 40	87 80 61 65 31	91 51 80 32 44
10 08 58 21 66	72 68 49 29 31	89 85 84 46 06	59 73 19 85 23	65 09 29 75 63
47 90 56 10 08	88 02 84 27 83	42 29 72 23 19	66 56 45 65 79	20 71 53 20 25
22 85 61 68 90	49 64 92 85 44	16 60 12 89 88	50 14 49 81 06	01 82 77 45 12
67 80 43 79 33	12 83 11 41 16	25 58 19 68 70	77 02 54 00 52	53 43 37 15 26
27 62 50 96 72	79 44 61 40 15	14 53 40 65 39	27 31 58 50 28	11 39 03 34 25
33 78 80 87 15	38 30 06 38 21	14 47 47 07 26	54 96 87 53 32	40 36 40 96 76
13 13 92 66 99	47 24 49 57 74	32 25 43 62 17	10 97 11 69 84	99 63 22 32 98

Table 37 Random Numbers (4) – continued

10 27 53 96 23	71 50 54 36 23	54 31 04 82 98	04 14 12 15 09	26 78 25 47 47
28 41 50 61 88	64 85 27 20 18	83 36 36 05 56	39 71 65 09 62	94 76 62 11 89
34 21 42 57 02	59 19 18 97 48	80 30 03 30 98	05 24 67 70 07	84 97 50 87 46
61 81 77 23 23	82 82 11 54 08	53 28 70 58 96	44 07 39 55 43	42 34 43 39 28
61 15 18 13 54	16 86 20 26 88	90 74 80 55 09	14 53 90 51 17	52 01 63 01 59
91 76 21 64 64	44 91 13 32 97	75 31 62 66 54	84 80 32 75 77	56 08 25 70 29
00 97 79 08 06	37 30 28 59 85	53 56 68 53 40	01 74 39 59 73	30 19 99 85 48
36 46 18 34 94	75 20 80 27 77	78 91 69 16 00	08 43 18 73 68	67 69 61 34 25
88 98 99 60 50	65 95 79 42 94	93 62 40 89 96	43 56 47 71 66	46 76 29 67 02
04 37 59 87 21	05 02 03 24 17	47 97 81 56 51	92 34 86 01 82	55 51 33 12 91
63 62 06 34 41	94 21 78 55 09	72 76 45 16 94	29 95 81 83 83	79 88 01 97 30
78 47 23 53 90	34 41 92 45 71	09 23 70 70 07	12 38 92 79 43	14 85 11 47 23
87 68 62 15 43	53 14 36 59 25	54 47 33 70 15	59 24 48 40 35	50 03 42 99 36
47 60 92 10 77	88 59 53 11 52	66 25 69 07 04	48 68 64 71 06	61 65 70 22 12
56 88 87 59 41	65 28 04 67 53	95 79 88 37 31	50 41 06 94 76	81 83 17 16 33
02 57 45 86 67	73 43 07 34 48	44 26 87 93 29	77 09 61 67 84	06 69 44 77 75
31 54 14 13 17	48 62 11 90 60	68 12 93 64 28	46 24 79 16 76	14 60 25 51 01
28 50 16 43 36	28 97 85 58 99	67 22 52 76 23	24 70 36 54 54	59 28 61 71 96
63 29 62 66 50	02 63 45 52 38	67 63 47 54 75	83 24 78 43 20	92 63 13 47 48
45 65 58 26 51	76 96 59 38 72	86 57 45 71 46	44 67 76 14 55	44 88 01 62 12
39 65 36 63 70	77 45 85 50 51	74 13 39 35 22	30 53 36 02 95	49 34 88 73 61
73 71 98 16 04	29 18 94 51 23	76 51 94 84 86	79 93 96 38 63	08 58 25 58 94
72 20 56 20 11	72 65 71 08 86	79 57 95 13 91	97 48 72 66 48	09 71 17 24 89
75 17 26 99 76	89 37 20 70 01	77 31 61 95 46	26 97 05 73 51	53 33 18 72 87
37 48 69 82 29	81 30 15 39 14	48 38 75 93 29	06 87 37 78 48	45 56 00 84 47
68 08 02 80 72	83 71 46 30 49	89 17 95 88 29	02 39 56 03 46	97 74 06 56 17
14 23 98 61 67	70 52 85 01 50	01 84 02 78 43	10 62 98 19 41	18 83 99 47 99
49 08 96 21 44	25 27 99 41 28	07 41 08 34 66	19 42 74 39 91	41 96 53 78 72
78 37 06 08 43	63 61 62 42 29	39 68 95 10 96	09 24 23 00 62	56 12 80 73 16
37 21 34 17 68	68 96 83 23 56	32 84 60 15 31	44 73 67 34 77	91 15 79 74 58
14 29 09 34 04	87 83 07 55 07	76 58 30 83 64	87 29 25 58 84	86 50 60 00 25
58 43 28 06 36	49 52 83 51 14	47 56 92 29 34	05 87 31 06 95	12 45 57 09 09
10 43 67 29 70	80 62 80 03 42	10 80 21 38 84	90 56 35 03 09	43 12 74 49 14
44 38 88 39 54	86 97 37 44 22	00 95 01 31 76	17 16 29 56 63	38 78 94 49 81
90 69 59 19 51	85 39 52 85 13	07 28 37 07 61	11 16 36 27 03	78 86 72 04 95
41 47 10 25 62	97 05 31 03 61	20 26 36 31 62	68 69 86 95 44	84 95 48 46 45
91 94 14 63 19	75 89 11 47 11	31 56 34 19 09	79 57 92 36 59	14 93 87 81 40
80 06 54 18 66	09 18 94 06 19	98 40 07 17 81	22 45 44 84 11	24 62 20 42 31
67 72 77 63 48	84 08 31 55 58	24 33 45 77 58	80 45 67 93 82	75 70 16 08 24
59 40 24 13 27	79 26 88 86 30	01 31 60 10 39	53 58 47 70 93	85 81 56 39 38
05 90 35 89 95	01 61 16 96 94	50 78 13 69 36	37 68 53 37 31	71 26 35 03 71
44 43 80 69 98	46 68 05 14 82	90 78 50 05 62	77 79 13 57 44	59 60 10 39 66
61 81 31 96 82	00 57 25 60 59	46 72 60 18 77	55 66 12 62 11	08 99 55 64 57
42 88 07 10 05	24 98 65 63 21	47 21 61 88 32	27 80 30 21 60	10 92 35 36 12
77 94 30 05 39	28 10 99 00 27	12 73 73 99 12	49 99 57 94 82	96 88 57 17 91
78 83 19 76 16	94 11 68 84 26	23 54 20 86 85	23 86 66 99 07	36 37 34 92 09
87 76 59 61 81	43 63 64 61 61	65 76 36 95 90	18 48 27 45 68	27 23 65 30 72
91 43 05 96 47	55 78 99 95 24	37 55 85 78 78	01 48 41 19 10	35 19 54 07 73
84 97 77 72 73	09 62 06 65 72	87 12 49 03 60	41 15 20 76 27	50 47 02 29 16
87 41 60 76 83	44 88 96 07 80	83 05 83 38 96	73 70 66 81 90	30 56 10 48 59

Table 37 Random Numbers (5) – continued

28 89 65 87 08	13 50 63 04 23	25 47 57 91 31	52 62 24 19 94	91 67 48 57 10
30 29 43 65 42	78 66 28 55 80	47 46 41 90 08	55 98 78 10 70	49 92 05 12 07
95 74 62 60 53	51 57 32 22 27	12 72 72 27 77	44 67 32 23 13	67 95 07 76 30
01 85 54 96 72	66 86 65 64 60	56 59 75 36 75	46 44 33 63 71	54 50 06 44 75
10 91 46 96 86	19 83 52 47 53	65 00 51 93 51	30 80 05 19 29	56 23 27 19 03
05 33 18 08 51	51 78 57 26 17	34 87 96 23 95	89 99 93 39 79	11 28 94 15 52
04 43 13 37 00	79 68 96 26 60	70 39 83 66 56	62 03 55 86 57	77 55 33 62 02
05 85 40 25 24	73 52 93 70 50	48 21 47 74 63	17 27 27 51 26	35 96 29 00 45
84 90 90 65 77	63 99 25 69 02	09 04 03 35 78	19 79 95 07 21	02 84 48 51 97
28 55 53 09 48	86 28 30 02 35	71 30 32 06 47	93 74 21 86 33	49 90 21 69 74
89 83 40 69 80	97 96 47 59 97	56 33 24 87 36	17 18 16 90 46	75 27 28 52 13
73 20 96 05 68	93 41 69 96 07	97 50 81 79 59	42 37 13 81 83	82 42 85 04 31
10 89 07 76 21	40 24 74 36 42	40 33 04 46 24	35 63 02 31 61	34 59 43 36 96
91 50 27 78 37	06 06 16 25 98	17 78 80 36 85	26 41 77 63 37	71 63 94 94 33
03 45 44 66 88	97 81 26 03 89	39 46 67 21 17	98 10 39 33 15	61 63 00 25 92
89 41 58 91 63	65 99 59 97 84	90 14 79 61 55	56 16 88 87 60	32 15 99 67 43
13 43 00 97 26	16 91 21 32 41	60 22 66 72 17	31 85 33 69 07	68 49 20 43 29
71 71 00 51 72	62 03 89 26 32	35 27 99 18 25	78 12 03 09 70	50 93 19 35 56
19 28 15 00 41	92 27 73 40 38	37 11 05 75 16	98 81 99 37 29	92 20 32 39 67
56 38 30 92 30	45 51 94 69 04	00 84 14 36 37	95 66 39 01 09	21 68 40 95 79
39 27 52 89 11	00 81 06 28 48	12 08 05 75 26	03 35 63 05 77	13 81 20 67 58
73 13 28 58 01	05 06 42 24 07	60 60 29 99 93	72 93 78 04 36	25 76 01 54 03
81 60 84 51 57	12 68 46 55 89	60 09 71 87 89	70 81 10 95 91	83 79 68 20 66
05 62 98 07 85	07 79 26 69 61	67 85 72 37 41	85 79 76 84 23	61 58 87 08 05
62 97 16 29 18	52 16 16 23 56	62 95 80 97 63	32 25 34 03 36	48 84 60 37 65
31 13 63 21 08	16 01 92 58 21	48 79 74 73 72	08 64 80 91 38	07 28 66 61 59
97 38 35 34 19	89 84 05 34 47	88 09 31 54 88	97 96 86 01 69	46 13 95 65 96
32 11 78 33 82	51 99 98 44 39	12 75 10 60 36	80 66 39 94 97	42 36 31 16 59
81 99 13 37 05	08 12 60 39 23	61 73 84 89 18	26 02 04 37 95	96 18 69 06 30
45 74 00 03 05	69 99 47 26 52	48 06 30 00 18	03 30 28 55 59	66 10 71 44 05
11 84 13 69 01	88 91 28 79 50	71 42 14 96 55	98 59 96 01 36	88 77 90 45 59
14 66 12 87 22	59 45 27 08 51	85 64 23 85 41	64 72 08 59 44	67 98 56 65 56
40 25 67 87 82	84 27 17 30 37	48 69 49 02 58	98 02 50 58 11	95 39 06 35 63
44 48 97 49 43	65 45 53 41 07	14 83 46 74 11	76 66 63 60 08	90 54 33 65 84
41 94 54 06 57	48 28 01 83 84	09 11 21 91 73	97 28 44 74 06	22 30 95 69 72
07 12 15 58 84	93 18 31 83 45	54 52 62 29 91	53 58 54 66 05	47 19 63 92 75
64 27 90 43 52	18 26 32 96 83	50 58 45 27 57	14 96 39 64 85	73 87 96 76 23
80 71 86 41 03	45 62 63 40 88	35 69 34 10 94	32 22 52 04 74	69 63 21 83 41
27 06 08 09 92	26 22 59 28 27	38 58 22 14 79	24 32 12 38 42	33 56 90 92 57
54 68 97 20 54	33 26 74 03 30	74 22 19 13 48	30 28 01 92 49	58 61 52 27 03
02 92 65 68 99	05 53 15 26 70	04 69 22 64 07	04 73 25 74 82	78 35 22 21 88
83 52 57 78 62	98 61 70 48 22	68 50 64 55 75	42 70 32 09 60	58 70 61 43 97
82 82 76 31 33	85 13 41 38 10	16 47 61 43 77	83 27 19 70 41	34 78 77 60 25
38 61 34 09 49	04 41 66 09 76	20 50 73 40 95	24 77 95 73 20	47 42 80 61 03
01 01 11 88 38	03 10 16 82 24	39 58 20 12 39	82 77 02 18 88	33 11 49 15 16
21 66 14 38 28	54 08 18 07 04	92 17 63 36 75	33 14 11 11 78	97 30 53 62 38
32 29 30 69 59	68 50 33 31 47	15 64 88 75 27	04 51 41 61 96	86 62 93 66 71
04 59 21 65 47	39 90 89 86 77	46 86 86 88 86	50 09 13 24 91	54 80 67 78 66
38 64 50 07 36	56 50 45 94 25	48 28 48 30 51	60 73 73 03 87	68 47 37 10 84
48 33 50 83 53	59 77 64 59 90	58 92 62 50 18	93 09 45 89 06	13 26 98 86 29

Table 38 Random Standardised Normal Deviates (*Z* Values)

The numbers in the table constitute a 'population' of standardised normal deviates arranged in a random sequence; they may be used where a small sample (that is, not greater than 50 and preferably less) of normal deviates is required. To do this, start at a random co-ordinate position in the table and read off the required number of values either down, up, right or left from this starting deviate.

An indefinitely large random sample of these 'preferred' z-values may be taken with replacement in conjunction with a table of random numbers. Take a three-digit random number, using the first two digits to select a row and the third to select a column; the intersection gives a random standardised normal deviate. For example, if the random number is 861, the remainder after dividing by 500 is 361 and so the corresponding normal deviate is + 0.539. (The device of mapping the random numbers 500 to 999 on to those from 000 to 499 saves 'wasting' random numbers.)

The mean of the 500 tabulated values is 0.00 and the variance is 1.00.

	0	1	2	3	4	5	6	7	8	9
00	−0.179	−0.399	−0.235	−0.098	−0.465	+1.563	−1.085	+0.860	+0.388	+0.710
01	+0.421	+1.454	+0.904	+0.437	−2.120	+1.085	−0.277	−0.170	+0.018	−0.722
02	+0.210	−0.556	+0.465	−1.812	−2.748	−0.345	−0.251	+0.622	−1.015	+0.762
03	−1.598	+0.919	−0.266	−0.999	+0.308	−0.592	+0.817	−0.454	+1.598	+0.240
04	+1.717	+1.514	−0.012	−0.852	+0.118	+0.399	−0.123	+0.432	−0.470	+0.776
05	−0.308	+0.867	−0.372	+0.697	−1.787	+0.568	−0.002	−0.133	+0.545	−0.824
06	−0.421	+0.516	−0.038	+1.200	+0.063	−0.377	−1.007	−0.334	+1.299	+0.038
07	−0.776	+0.874	−1.265	−0.580	+0.377	−0.697	−2.226	−1.299	−0.796	−0.628
08	+0.640	−0.522	+0.023	−0.393	−1.412	−2.457	−1.580	+1.160	+0.008	+0.487
09	−0.319	+0.889	+1.180	−0.404	+1.322	+0.410	+1.468	+0.235	−0.810	−1.131
10	+0.610	−0.383	+1.812	+0.729	+0.204	−0.225	+0.169	−0.729	−0.432	+0.634
11	−0.174	−0.154	+0.098	+0.393	−3.090	+1.762	+1.530	+0.028	+0.950	−0.935
12	+2.576	−0.684	−1.200	+0.002	+0.261	−0.415	+0.598	−0.769	−0.169	−1.498
13	−1.103	+1.398	−0.653	+1.739	+0.476	+0.510	+0.782	−0.634	+0.562	−0.053
14	+1.635	+0.448	−1.530	−0.043	+2.290	−0.063	−1.695	+0.199	+1.211	−1.360
15	−0.068	−0.860	−0.194	−1.616	+0.334	+0.189	+0.927	−1.454	+0.958	+0.404
16	−1.960	+1.076	−0.671	−0.103	+1.041	+2.226	+1.838	−0.510	−1.322	+2.366
17	+0.443	−0.912	+0.251	−0.574	+1.131	−0.204	−0.324	−0.487	−1.287	+0.522
18	+1.360	+0.533	+1.094	+0.671	+0.852	−2.576	−0.539	−0.568	+0.225	−0.545
19	+0.810	+0.319	−1.514	+0.556	+1.112	−0.210	+0.292	+0.749	+0.882	+0.033
20	+0.616	+1.347	−1.866	−0.755	+0.329	+0.148	−0.058	−0.199	+0.048	+1.546
21	−0.598	−2.366	−0.831	+0.454	−0.118	−1.762	+0.493	+1.103	+0.361	+0.113
22	+0.426	+1.580	−1.112	+0.550	−1.254	−0.033	+0.143	−1.141	+0.366	−0.073
23	+0.831	−0.516	−1.717	−0.340	+1.655	+0.194	−0.388	−0.942	−1.243	−0.292
24	−0.640	−0.128	+1.276	−1.838	−0.410	+0.646	+2.075	−0.159	+1.695	+0.527
25	−0.927	+0.838	−1.546	+0.246	−0.742	−0.143	+2.457	+0.043	−1.058	−0.867
26	+1.232	+2.170	+0.088	−0.803	+0.574	+0.058	+0.282	+0.356	+0.350	−1.927
27	+0.935	+0.665	+2.034	−1.995	+0.703	−0.083	−1.468	+0.078	−0.966	−0.303
28	−1.739	−0.622	−1.563	+0.313	+0.220	−0.586	+0.272	+0.789	−1.335	+1.440
29	+0.990	−1.483	+0.154	−1.372	−1.896	+1.385	−1.041	+0.974	+0.482	−1.211
30	−0.189	−0.240	+0.133	−2.290	−0.616	−0.437	+0.459	−0.499	+0.845	+0.383
31	+1.866	−1.398	+0.068	+0.053	−2.034	+1.426	+1.254	+1.067	+0.592	+0.174
32	−0.018	+0.628	+0.230	+0.659	−0.298	+1.927	−0.282	+0.769	−0.690	+1.675
33	−0.646	−0.350	+0.324	−1.675	+1.190	−1.076	+1.287	−1.426	+0.345	−0.215
34	−1.150	−0.220	−0.533	+0.912	−0.710	−0.904	−0.817	−1.160	−0.919	−0.659
35	+0.103	−0.361	+1.024	−0.604	+0.966	−1.122	+0.604	−0.845	+0.736	−0.882
36	+1.243	+0.539	+0.684	−0.716	−0.482	−0.562	+0.277	−1.440	−0.366	−0.256
37	−0.093	−1.190	+0.580	−1.276	+0.563	−0.048	+0.742	−1.170	+1.960	+2.120
38	−0.261	−0.194	+0.303	+0.340	+1.498	−1.232	−0.078	−0.443	+1.141	+1.787
39	−0.230	−0.550	+0.266	−1.655	+0.999	−1.067	+1.058	+0.796	+0.415	+1.995
40	−0.148	+0.504	−0.028	+0.083	+0.824	−1.024	+1.412	−0.164	+1.150	−0.272
41	+1.122	+0.896	−0.789	+0.215	−0.426	−1.049	−0.974	+0.586	+1.311	−0.736
42	+0.499	−1.032	+0.159	+0.123	+2.748	−0.749	−0.665	−1.221	−1.180	+1.049
43	+0.678	−0.782	+0.470	+0.256	+0.298	−0.990	+0.287	+0.942	+0.128	+1.372
44	−1.347	+3.090	−0.896	+0.138	−0.838	+0.690	+1.007	+0.184	+0.164	+0.179
45	−1.094	−0.610	−0.287	+0.755	−0.459	−1.635	−0.108	−0.246	+1.032	−0.527
46	−0.088	−0.889	+0.803	−1.311	−0.703	+1.170	−0.113	+0.108	−0.874	+0.372
47	+0.093	−0.476	+1.265	−0.448	+1.015	−0.313	−0.958	+0.716	+1.483	+0.722
48	−0.950	−0.008	+0.012	+0.073	−0.762	−0.493	+1.896	+0.982	+1.616	+1.221
49	−0.329	−0.138	−0.504	−0.678	+1.335	−2.075	−1.385	−0.023	−0.356	−0.982

Table 39 Present Value Factors

The table gives the present value of a single payment received *n* years in the future discounted at *x*% per year.
For example, with a discount rate of 7%, a single payment of £1 at the end of six years has a present value of £0.6663 or 66.63p.
The table can also be used to give the value of an investment which grows by compound interest of *x*% per year for *n* years.
For example, £1 invested at 7% compounded for 6 years would be worth £1/ 0.6663 = £1.500

Year	1%	2%	3%	4%	5%	6%	7%	8%	9%	10%
1	.9901	.9804	.9709	.9615	.9524	.9434	.9346	.9259	.9174	.9091
2	.9803	.9612	.9426	.9246	.9070	.8900	.8734	.8573	.8417	.8264
3	.9706	.9423	.9151	.8890	.8638	.8396	.8163	.7938	.7722	.7513
4	.9610	.9238	.8885	.8548	.8227	.7921	.7629	.7350	.7084	.6830
5	.9515	.9057	.8626	.8219	.7835	.7473	.7130	.6806	.6499	.6209
6	.9420	.8880	.8375	.7903	.7462	.7050	.6663	.6302	.5963	.5645
7	.9327	.8706	.8131	.7599	.7107	.6651	.6227	.5835	.5470	.5132
8	.9235	.8535	.7894	.7307	.6768	.6274	.5820	.5403	.5019	.4665
9	.9143	.8368	.7664	.7026	.6446	.5919	.5439	.5002	.4604	.4241
10	.9053	.8203	.7441	.6756	.6139	.5584	.5083	.4632	.4224	.3855
11	.8963	.8043	.7224	.6496	.5847	.5268	.4751	.4289	.3875	.3505
12	.8874	.7885	.7014	.6246	.5568	.4970	.4440	.3971	.3555	.3186
13	.8787	.7730	.6810	.6006	.5303	.4688	.4150	.3677	.3262	.2897
14	.8700	.7579	.6611	.5775	.5051	.4423	.3878	.3405	.2992	.2633
15	.8613	.7430	.6419	.5553	.4810	.4173	.3624	.3152	.2745	.2394
16	.8528	.7284	.6232	.5339	.4581	.3936	.3387	.2919	.2519	.2176
17	.8444	.7142	.6050	.5134	.4363	.3714	.3166	.2703	.2311	.1978
18	.8360	.7002	.5874	.4936	.4155	.3503	.2959	.2502	.2120	.1799
19	.8277	.6864	.5703	.4746	.3957	.3305	.2765	.2317	.1945	.1635
20	.8195	.6730	.5537	.4564	.3769	.3118	.2584	.2145	.1784	.1486
21	.8114	.6598	.5375	.4388	.3589	.2942	.2415	.1987	.1637	.1351
22	.8034	.6468	.5219	.4220	.3418	.2775	.2257	.1839	.1502	.1228
23	.7954	.6342	.5067	.4057	.3256	.2618	.2109	.1703	.1378	.1117
24	.7876	.6217	.4919	.3901	.3101	.2470	.1971	.1577	.1264	.1015
25	.7798	.6095	.4776	.3751	.2953	.2330	.1842	.1460	.1160	.0923
26	.7720	.5976	.4637	.3607	.2812	.2198	.1722	.1352	.1064	.0839
27	.7644	.5859	.4502	.3468	.2678	.2074	.1609	.1252	.0976	.0763
28	.7568	.5744	.4371	.3335	.2551	.1956	.1504	.1159	.0895	.0693
29	.7493	.5631	.4243	.3207	.2429	.1846	.1406	.1073	.0822	.0630
30	.7419	.5521	.4120	.3083	.2314	.1741	.1314	.0994	.0754	.0573
31	.7346	.5412	.4000	.2965	.2204	.1643	.1228	.0920	.0691	.0521
32	.7273	.5306	.3883	.2851	.2099	.1550	.1147	.0852	.0634	.0474
33	.7201	.5202	.3770	.2741	.1999	.1462	.1072	.0789	.0582	.0431
34	.7130	.5100	.3660	.2636	.1904	.1379	.1002	.0730	.0534	.0391
35	.7059	.5000	.3554	.2534	.1813	.1301	.0937	.0676	.0490	.0356
36	.6989	.4902	.3450	.2437	.1727	.1227	.0875	.0626	.0449	.0323
37	.6920	.4806	.3350	.2343	.1644	.1158	.0818	.0580	.0412	.0294
38	.6852	.4712	.3252	.2253	.1566	.1092	.0765	.0537	.0378	.0267
39	.6784	.4619	.3158	.2166	.1491	.1031	.0715	.0497	.0347	.0243
40	.6717	.4529	.3066	.2083	.1420	.0972	.0668	.0460	.0318	.0221
41	.6650	.4440	.2976	.2003	.1353	.0917	.0624	.0426	.0292	.0201
42	.6584	.4353	.2890	.1926	.1288	.0865	.0583	.0395	.0268	.0183
43	.6519	.4268	.2805	.1852	.1227	.0816	.0545	.0365	.0246	.0166
44	.6454	.4184	.2724	.1780	.1169	.0770	.0509	.0338	.0226	.0151
45	.6391	.4102	.2644	.1712	.1113	.0727	.0476	.0313	.0207	.0137
46	.6327	.4022	.2567	.1646	.1060	.0685	.0445	.0290	.0190	.0125
47	.6265	.3943	.2493	.1583	.1009	.0647	.0416	.0269	.0174	.0113
48	.6203	.3865	.2420	.1522	.0961	.0610	.0389	.0249	.0160	.0103
49	.6141	.3790	.2350	.1463	.0916	.0575	.0363	.0230	.0147	.0094
50	.6080	.3715	.2281	.1407	.0872	.0543	.0339	.0213	.0134	.0085

Table 39 Present Value Factors – continued

Discount range 11% to 20%

Year / x	11%	12%	13%	14%	15%	16%	17%	18%	19%	20%
1	.9009	.8929	.8850	.8772	.8696	.8621	.8547	.8475	.8403	.8333
2	.8116	.7972	.7831	.7695	.7561	.7432	.7305	.7182	.7062	.6944
3	.7312	.7118	.6931	.6750	.6575	.6407	.6244	.6086	.5934	.5787
4	.6587	.6355	.6133	.5921	.5718	.5523	.5337	.5158	.4987	.4823
5	.5935	.5674	.5428	.5194	.4972	.4761	.4561	.4371	.4190	.4019
6	.5346	.5066	.4803	.4556	.4323	.4104	.3898	.3704	.3521	.3349
7	.4817	.4523	.4251	.3996	.3759	.3538	.3332	.3139	.2959	.2791
8	.4339	.4039	.3762	.3506	.3269	.3050	.2848	.2660	.2487	.2326
9	.3909	.3606	.3329	.3075	.2843	.2630	.2434	.2255	.2090	.1938
10	.3522	.3220	.2946	.2697	.2472	.2267	.2080	.1911	.1756	.1615
11	.3173	.2875	.2607	.2366	.2149	.1954	.1778	.1619	.1476	.1346
12	.2858	.2567	.2307	.2076	.1869	.1685	.1520	.1372	.1240	.1122
13	.2575	.2292	.2042	.1821	.1625	.1452	.1299	.1163	.1042	.0935
14	.2320	.2046	.1807	.1597	.1413	.1252	.1110	.0985	.0876	.0779
15	.2090	.1827	.1599	.1401	.1229	.1079	.0949	.0835	.0736	.0649
16	.1883	.1631	.1415	.1229	.1069	.0930	.0811	.0708	.0618	.0541
17	.1696	.1456	.1252	.1078	.0929	.0802	.0693	.0600	.0520	.0451
18	.1528	.1300	.1108	.0946	.0808	.0691	.0592	.0508	.0437	.0376
19	.1377	.1161	.0981	.0829	.0703	.0596	.0506	.0431	.0367	.0313
20	.1240	.1037	.0868	.0728	.0611	.0514	.0433	.0365	.0308	.0261
21	.1117	.0926	.0768	.0638	.0531	.0443	.0370	.0309	.0259	.0217
22	.1007	.0826	.0680	.0560	.0462	.0382	.0316	.0262	.0218	.0181
23	.0907	.0738	.0601	.0491	.0402	.0329	.0270	.0222	.0183	.0151
24	.0817	.0659	.0532	.0431	.0349	.0284	.0231	.0188	.0154	.0126
25	.0736	.0588	.0471	.0378	.0304	.0245	.0197	.0160	.0129	.0105
26	.0663	.0525	.0417	.0331	.0264	.0211	.0169	.0135	.0109	.0087
27	.0597	.0469	.0369	.0291	.0230	.0182	.0144	.0115	.0091	.0073
28	.0538	.0419	.0326	.0255	.0200	.0157	.0123	.0097	.0077	.0061
29	.0485	.0374	.0289	.0224	.0174	.0135	.0105	.0082	.0064	.0051
30	.0437	.0334	.0256	.0196	.0151	.0116	.0090	.0070	.0054	.0042
31	.0394	.0298	.0226	.0172	.0131	.0100	.0077	.0059	.0046	.0035
32	.0355	.0266	.0200	.0151	.0114	.0087	.0066	.0050	.0038	.0029
33	.0319	.0238	.0177	.0132	.0099	.0075	.0056	.0042	.0032	.0024
34	.0288	.0212	.0157	.0116	.0086	.0064	.0048	.0036	.0027	.0020
35	.0259	.0189	.0139	.0102	.0075	.0055	.0041	.0030	.0023	.0017
36	.0234	.0169	.0123	.0089	.0065	.0048	.0035	.0026	.0019	.0014
37	.0210	.0151	.0109	.0078	.0057	.0041	.0030	.0022	.0016	.0012
38	.0190	.0135	.0096	.0069	.0049	.0036	.0026	.0019	.0013	.0010
39	.0171	.0120	.0085	.0060	.0043	.0031	.0022	.0016	.0011	.0008
40	.0154	.0107	.0075	.0053	.0037	.0026	.0019	.0013	.0010	.0007
41	.0139	.0096	.0067	.0046	.0032	.0023	.0016	.0011	.0008	.0006
42	.0125	.0086	.0059	.0041	.0028	.0020	.0014	.0010	.0007	.0005
43	.0112	.0076	.0052	.0036	.0025	.0017	.0012	.0008	.0006	.0004
44	.0101	.0068	.0046	.0031	.0021	.0015	.0010	.0007	.0005	.0003
45	.0091	.0061	.0041	.0027	.0019	.0013	.0009	.0006	.0004	.0003
46	.0082	.0054	.0036	.0024	.0016	.0011	.0007	.0005	.0003	.0002
47	.0074	.0049	.0032	.0021	.0014	.0009	.0006	.0004	.0003	.0002
48	.0067	.0043	.0028	.0019	.0012	.0008	.0005	.0004	.0002	.0002
49	.0060	.0039	.0025	.0016	.0011	.0007	.0005	.0003	.0002	.0001
50	.0054	.0035	.0022	.0014	.0009	.0006	.0004	.0003	.0002	.0001

Table 39 Present Value Factors – continued

Discount range 21% to 30%

Year (x)	21%	22%	23%	24%	25%	26%	27%	28%	29%	30%
1	0.8264	0.8197	0.8130	0.8065	0.8000	0.7937	0.7874	0.7813	0.7752	0.7692
2	0.6830	0.6719	0.6610	0.6504	0.6400	0.6299	0.6200	0.6104	0.6009	0.5917
3	0.5645	0.5507	0.5374	0.5245	0.5120	0.4999	0.4882	0.4768	0.4658	0.4552
4	0.4665	0.4514	0.4369	0.4230	0.4096	0.3968	0.3844	0.3725	0.3611	0.3501
5	0.3855	0.3700	0.3552	0.3411	0.3277	0.3149	0.3027	0.2910	0.2799	0.2693
6	0.3186	0.3033	0.2888	0.2751	0.2621	0.2499	0.2383	0.2274	0.2170	0.2072
7	0.2633	0.2486	0.2348	0.2218	0.2097	0.1983	0.1877	0.1776	0.1682	0.1594
8	0.2176	0.2038	0.1909	0.1789	0.1678	0.1574	0.1478	0.1388	0.1304	0.1226
9	0.1799	0.1670	0.1552	0.1443	0.1342	0.1249	0.1164	0.1084	0.1011	0.0943
10	0.1486	0.1369	0.1262	0.1164	0.1074	0.0992	0.0916	0.0847	0.0784	0.0725
11	0.1228	0.1122	0.1026	0.0938	0.0859	0.0787	0.0721	0.0662	0.0607	0.0558
12	0.1015	0.0920	0.0834	0.0757	0.0687	0.0625	0.0568	0.0517	0.0471	0.0429
13	0.0839	0.0754	0.0678	0.0610	0.0550	0.0496	0.0447	0.0404	0.0365	0.0330
14	0.0693	0.0618	0.0551	0.0492	0.0440	0.0393	0.0352	0.0316	0.0283	0.0254
15	0.0573	0.0507	0.0448	0.0397	0.0352	0.0312	0.0277	0.0247	0.0219	0.0195
16	0.0474	0.0415	0.0364	0.0320	0.0281	0.0248	0.0218	0.0193	0.0170	0.0150
17	0.0391	0.0340	0.0296	0.0258	0.0225	0.0197	0.0172	0.0150	0.0132	0.0116
18	0.0323	0.0279	0.0241	0.0208	0.0180	0.0156	0.0135	0.0118	0.0102	0.0089
19	0.0267	0.0229	0.0196	0.0168	0.0144	0.0124	0.0107	0.0092	0.0079	0.0068
20	0.0221	0.0187	0.0159	0.0135	0.0115	0.0098	0.0084	0.0072	0.0061	0.0053
21	0.0183	0.0154	0.0129	0.0109	0.0092	0.0078	0.0066	0.0056	0.0048	0.0040
22	0.0151	0.0126	0.0105	0.0088	0.0074	0.0062	0.0052	0.0044	0.0037	0.0031
23	0.0125	0.0103	0.0086	0.0071	0.0059	0.0049	0.0041	0.0034	0.0029	0.0024
24	0.0103	0.0085	0.0070	0.0057	0.0047	0.0039	0.0032	0.0027	0.0022	0.0018
25	0.0085	0.0069	0.0057	0.0046	0.0038	0.0031	0.0025	0.0021	0.0017	0.0014
26	0.0070	0.0057	0.0046	0.0037	0.0030	0.0025	0.0020	0.0016	0.0013	0.0011
27	0.0058	0.0047	0.0037	0.0030	0.0024	0.0019	0.0016	0.0013	0.0010	0.0008
28	0.0048	0.0038	0.0030	0.0024	0.0019	0.0015	0.0012	0.0010	0.0008	0.0006
29	0.0040	0.0031	0.0025	0.0020	0.0015	0.0012	0.0010	0.0008	0.0006	0.0005
30	0.0033	0.0026	0.0020	0.0016	0.0012	0.0010	0.0008	0.0006	0.0005	0.0004
31	0.0027	0.0021	0.0016	0.0013	0.0010	0.0008	0.0006	0.0005	0.0004	0.0003
32	0.0022	0.0017	0.0013	0.0010	0.0008	0.0006	0.0005	0.0004	0.0003	0.0002
33	0.0019	0.0014	0.0011	0.0008	0.0006	0.0005	0.0004	0.0003	0.0002	0.0002
34	0.0015	0.0012	0.0009	0.0007	0.0005	0.0004	0.0003	0.0002	0.0002	0.0001
35	0.0013	0.0009	0.0007	0.0005	0.0004	0.0003	0.0002	0.0002	0.0001	0.0001
36	0.0010	0.0008	0.0006	0.0004	0.0003	0.0002	0.0002	0.0001	0.0001	0.0001
37	0.0009	0.0006	0.0005	0.0003	0.0003	0.0002	0.0001	0.0001	0.0001	0.0001
38	0.0007	0.0005	0.0004	0.0003	0.0002	0.0002	0.0001	0.0001	0.0001	
39	0.0006	0.0004	0.0003	0.0002	0.0002	0.0001	0.0001	0.0001		
40	0.0005	0.0004	0.0003	0.0002	0.0001	0.0001	0.0001	0.0001		
41	0.0004	0.0003	0.0002	0.0001	0.0001	0.0001	0.0001			
42	0.0003	0.0002	0.0002	0.0001	0.0001	0.0001				
43	0.0003	0.0002	0.0001	0.0001	0.0001					
44	0.0002	0.0002	0.0001	0.0001	0.0001					
45	0.0002	0.0001	0.0001	0.0001						
46	0.0002	0.0001	0.0001	0.0001						
47	0.0001	0.0001	0.0001							
48	0.0001	0.0001								
49	0.0001	0.0001								
50	0.0001									

Table 39 Present Value Factors – continued

Discount range 31% to 40%

Year / x	31%	32%	33%	34%	35%	36%	37%	38%	39%	40%
1	0.7634	0.7576	0.7519	0.7463	0.7407	0.7353	0.7299	0.7246	0.7194	0.7143
2	0.5827	0.5739	0.5653	0.5569	0.5487	0.5407	0.5328	0.5251	0.5176	0.5102
3	0.4448	0.4348	0.4251	0.4156	0.4064	0.3975	0.3889	0.3805	0.3724	0.3644
4	0.3396	0.3294	0.3196	0.3102	0.3011	0.2923	0.2839	0.2757	0.2679	0.2603
5	0.2592	0.2495	0.2403	0.2315	0.2230	0.2149	0.2072	0.1998	0.1927	0.1859
6	0.1979	0.1890	0.1807	0.1727	0.1652	0.1580	0.1512	0.1448	0.1386	0.1328
7	0.1510	0.1432	0.1358	0.1289	0.1224	0.1162	0.1104	0.1049	0.0997	0.0949
8	0.1153	0.1085	0.1021	0.0962	0.0906	0.0854	0.0806	0.0760	0.0718	0.0678
9	0.0880	0.0822	0.0768	0.0718	0.0671	0.0628	0.0588	0.0551	0.0516	0.0484
10	0.0672	0.0623	0.0577	0.0536	0.0497	0.0462	0.0429	0.0399	0.0371	0.0346
11	0.0513	0.0472	0.0434	0.0400	0.0368	0.0340	0.0313	0.0289	0.0267	0.0247
12	0.0392	0.0357	0.0326	0.0298	0.0273	0.0250	0.0229	0.0210	0.0192	0.0176
13	0.0299	0.0271	0.0245	0.0223	0.0202	0.0184	0.0167	0.0152	0.0138	0.0126
14	0.0228	0.0205	0.0185	0.0166	0.0150	0.0135	0.0122	0.0110	0.0099	0.0090
15	0.0174	0.0155	0.0139	0.0124	0.0111	0.0099	0.0089	0.0080	0.0072	0.0064
16	0.0133	0.0118	0.0104	0.0093	0.0082	0.0073	0.0065	0.0058	0.0051	0.0046
17	0.0101	0.0089	0.0078	0.0069	0.0061	0.0054	0.0047	0.0042	0.0037	0.0033
18	0.0077	0.0068	0.0059	0.0052	0.0045	0.0039	0.0035	0.0030	0.0027	0.0023
19	0.0059	0.0051	0.0044	0.0038	0.0033	0.0029	0.0025	0.0022	0.0019	0.0017
20	0.0045	0.0039	0.0033	0.0029	0.0025	0.0021	0.0018	0.0016	0.0014	0.0012
21	0.0034	0.0029	0.0025	0.0021	0.0018	0.0016	0.0013	0.0012	0.0010	0.0009
22	0.0026	0.0022	0.0019	0.0016	0.0014	0.0012	0.0010	0.0008	0.0007	0.0006
23	0.0020	0.0017	0.0014	0.0012	0.0010	0.0008	0.0007	0.0006	0.0005	0.0004
24	0.0015	0.0013	0.0011	0.0009	0.0007	0.0006	0.0005	0.0004	0.0004	0.0003
25	0.0012	0.0010	0.0008	0.0007	0.0006	0.0005	0.0004	0.0003	0.0003	0.0002
26	0.0009	0.0007	0.0006	0.0005	0.0004	0.0003	0.0003	0.0002	0.0002	0.0002
27	0.0007	0.0006	0.0005	0.0004	0.0003	0.0002	0.0002	0.0002	0.0001	0.0001
28	0.0005	0.0004	0.0003	0.0003	0.0002	0.0002	0.0001	0.0001	0.0001	0.0001
29	0.0004	0.0003	0.0003	0.0002	0.0002	0.0001	0.0001	0.0001	0.0001	0.0001
30	0.0003	0.0002	0.0002	0.0002	0.0001	0.0001	0.0001	0.0001	0.0001	
31	0.0002	0.0002	0.0001	0.0001	0.0001	0.0001	0.0001			
32	0.0002	0.0001	0.0001	0.0001	0.0001	0.0001				
33	0.0001	0.0001	0.0001	0.0001	0.0001					
34	0.0001	0.0001	0.0001							
35	0.0001	0.0001								
36	0.0001									
37										
38										
39										
40										
41										
42										
43										
44										
45										
46										
47										
48										
49										
50										

Table 40 Cumulative Present Value Factors

The table gives the present value of *n* annual payments of £1 received at the end of each of the next *n* years with a constant discount rate of *x*%.
For example, with a discount rate of 7% and with 6 annual payments of £1 at the end of each year, the present value is £4.767.

Year	1%	2%	3%	4%	5%	6%	7%	8%	9%	10%
1	0.990	0.980	0.971	0.962	0.952	0.943	0.935	0.926	0.917	0.909
2	1.970	1.942	1.914	1.886	1.859	1.833	1.808	1.783	1.759	1.736
3	2.941	2.884	2.829	2.775	2.723	2.673	2.624	2.577	2.531	2.487
4	3.902	3.808	3.717	3.630	3.546	3.465	3.387	3.312	3.240	3.170
5	4.854	4.713	4.580	4.452	4.329	4.212	4.100	3.993	3.890	3.791
6	5.796	5.601	5.417	5.242	5.076	4.917	4.767	4.623	4.486	4.355
7	6.728	6.472	6.230	6.002	5.786	5.583	5.389	5.206	5.033	4.868
8	7.652	7.326	7.020	6.733	6.463	6.210	5.971	5.747	5.535	5.335
9	8.566	8.162	7.786	7.435	7.108	6.802	6.515	6.247	5.995	5.759
10	9.471	8.983	8.530	8.111	7.722	7.360	7.023	6.710	6.418	6.145
11	10.368	9.787	9.253	8.761	8.306	7.887	7.499	7.139	6.805	6.495
12	11.255	10.575	9.954	9.385	8.863	8.384	7.943	7.536	7.161	6.814
13	12.134	11.348	10.635	9.986	9.393	8.853	8.358	7.904	7.487	7.103
14	13.004	12.106	11.296	10.563	9.899	9.295	8.745	8.244	7.786	7.367
15	13.865	12.849	11.938	11.119	10.380	9.712	9.108	8.559	8.061	7.606
16	14.718	13.578	12.561	11.652	10.838	10.106	9.446	8.851	8.312	7.824
17	15.562	14.292	13.166	12.166	11.274	10.477	9.763	9.122	8.544	8.021
18	16.398	14.992	13.754	12.659	11.689	10.828	10.059	9.372	8.756	8.201
19	17.226	15.679	14.324	13.134	12.085	11.158	10.335	9.604	8.950	8.365
20	18.045	16.352	14.878	13.590	12.462	11.470	10.594	9.818	9.128	8.513
21	18.857	17.011	15.415	14.029	12.821	11.764	10.835	10.017	9.292	8.649
22	19.660	17.658	15.937	14.451	13.163	12.042	11.061	10.201	9.442	8.771
23	20.456	18.292	16.444	14.857	13.488	12.304	11.272	10.371	9.580	8.883
24	21.243	18.914	16.936	15.247	13.798	12.551	11.469	10.529	9.707	8.985
25	22.023	19.524	17.413	15.622	14.094	12.784	11.653	10.675	9.823	9.077
26	22.795	20.121	17.877	15.983	14.375	13.003	11.825	10.810	9.929	9.161
27	23.559	20.707	18.327	16.330	14.643	13.211	11.986	10.935	10.027	9.237
28	24.316	21.281	18.764	16.663	14.898	13.406	12.137	11.051	10.116	9.306
29	25.066	21.845	19.189	16.984	15.141	13.591	12.277	11.158	10.198	9.369
30	25.807	22.397	19.601	17.292	15.372	13.765	12.409	11.258	10.274	9.427
31	26.542	22.938	20.001	17.589	15.593	13.929	12.532	11.350	10.343	9.479
32	27.269	23.468	20.389	17.874	15.802	14.084	12.646	11.435	10.406	9.526
33	27.989	23.989	20.766	18.148	16.002	14.231	12.753	11.514	10.464	9.569
34	28.702	24.499	21.132	18.411	16.193	14.368	12.854	11.587	10.518	9.608
35	29.408	24.999	21.487	18.665	16.374	14.499	12.947	11.654	10.567	9.644
36	30.107	25.489	21.832	18.909	16.547	14.621	13.035	11.717	10.612	9.676
37	30.799	25.969	22.167	19.143	16.711	14.737	13.117	11.775	10.653	9.706
38	31.484	26.441	22.492	19.368	16.868	14.846	13.193	11.829	10.691	9.732
39	32.163	26.903	22.808	19.585	17.017	14.949	13.265	11.878	10.725	9.757
40	32.835	27.355	23.115	19.793	17.159	15.047	13.331	11.924	10.757	9.779
41	33.500	27.799	23.412	19.993	17.294	15.138	13.394	11.967	10.786	9.799
42	34.158	28.235	23.701	20.186	17.423	15.225	13.452	12.006	10.813	9.817
43	34.810	28.662	23.982	20.371	17.546	15.306	13.507	12.043	10.838	9.834
44	35.455	29.080	24.254	20.549	17.663	15.383	13.558	12.077	10.860	9.849
45	36.094	29.490	24.519	20.720	17.774	15.456	13.605	12.108	10.881	9.863
46	36.727	29.892	24.775	20.885	17.880	15.525	13.650	12.137	10.900	9.875
47	37.354	30.287	25.025	21.043	17.981	15.589	13.691	12.164	10.917	9.886
48	37.974	30.673	25.267	21.195	18.077	15.650	13.730	12.189	10.933	9.897
49	38.588	31.052	25.502	21.342	18.168	15.708	13.766	12.212	10.948	9.906
50	39.196	31.424	25.730	21.482	18.256	15.762	13.800	12.233	10.962	9.915

Table 40 Cumulative Present Value Factors – continued

Discount range 11% to 20%.

x / Year	11%	12%	13%	14%	15%	16%	17%	18%	19%	20%
1	0.901	0.893	0.885	0.877	0.870	0.862	0.855	0.848	0.840	0.833
2	1.713	1.690	1.668	1.647	1.626	1.605	1.585	1.566	1.547	1.528
3	2.444	2.402	2.361	2.322	2.283	2.246	2.210	2.174	2.140	2.106
4	3.102	3.037	2.975	2.914	2.855	2.798	2.743	2.690	2.639	2.589
5	3.696	3.605	3.517	3.433	3.352	3.274	3.199	3.127	3.058	2.991
6	4.231	4.111	3.998	3.889	3.785	3.685	3.589	3.498	3.410	3.326
7	4.712	4.564	4.423	4.288	4.160	4.039	3.922	3.812	3.706	3.605
8	5.146	4.968	4.799	4.639	4.487	4.344	4.207	4.078	3.954	3.837
9	5.537	5.328	5.132	4.947	4.772	4.607	4.451	4.303	4.163	4.031
10	5.889	5.650	5.426	5.216	5.019	4.833	4.659	4.494	4.339	4.193
11	6.207	5.938	5.687	5.453	5.234	5.029	4.836	4.656	4.487	4.327
12	6.492	6.194	5.918	5.660	5.421	5.197	4.988	4.793	4.611	4.439
13	6.750	6.424	6.122	5.843	5.583	5.342	5.118	4.910	4.715	4.533
14	6.982	6.628	6.303	6.002	5.724	5.468	5.229	5.008	4.802	4.611
15	7.191	6.811	6.463	6.142	5.847	5.576	5.324	5.092	4.876	4.676
16	7.379	6.974	6.604	6.265	5.954	5.669	5.405	5.162	4.938	4.730
17	7.549	7.120	6.729	6.373	6.047	5.749	5.475	5.222	4.990	4.775
18	7.702	7.250	6.840	6.468	6.128	5.818	5.534	5.273	5.033	4.812
19	7.839	7.366	6.938	6.551	6.198	5.877	5.584	5.316	5.070	4.844
20	7.963	7.469	7.025	6.623	6.259	5.929	5.628	5.353	5.101	4.870
21	8.075	7.562	7.102	6.687	6.312	5.973	5.665	5.384	5.127	4.892
22	8.176	7.645	7.170	6.743	6.359	6.011	5.696	5.410	5.149	4.910
23	8.266	7.718	7.230	6.792	6.399	6.044	5.723	5.432	5.167	4.925
24	8.348	7.784	7.283	6.835	6.434	6.073	5.746	5.451	5.182	4.937
25	8.422	7.843	7.330	6.873	6.464	6.097	5.766	5.467	5.195	4.948
26	8.488	7.896	7.372	6.906	6.491	6.118	5.783	5.480	5.206	4.957
27	8.548	7.943	7.409	6.935	6.514	6.136	5.797	5.492	5.215	4.964
28	8.601	7.984	7.441	6.961	6.534	6.152	5.810	5.502	5.223	4.970
29	8.650	8.022	7.470	6.983	6.551	6.166	5.820	5.510	5.229	4.975
30	8.694	8.055	7.496	7.003	6.566	6.177	5.829	5.517	5.235	4.979
31	8.733	8.085	7.519	7.020	6.579	6.187	5.837	5.523	5.239	4.983
32	8.769	8.112	7.539	7.035	6.591	6.196	5.844	5.528	5.243	4.986
33	8.800	8.135	7.556	7.048	6.600	6.203	5.849	5.532	5.246	4.988
34	8.829	8.157	7.572	7.060	6.609	6.210	5.854	5.535	5.249	4.990
35	8.855	8.176	7.586	7.070	6.617	6.215	5.858	5.538	5.251	4.992
36	8.879	8.192	7.598	7.079	6.623	6.220	5.862	5.541	5.253	4.993
37	8.900	8.208	7.609	7.087	6.629	6.224	5.865	5.543	5.255	4.994
38	8.919	8.221	7.619	7.094	6.634	6.228	5.867	5.545	5.256	4.995
39	8.936	8.233	7.627	7.100	6.638	6.231	5.869	5.547	5.257	4.996
40	8.951	8.244	7.635	7.105	6.642	6.234	5.871	5.548	5.258	4.997
41	8.965	8.253	7.641	7.110	6.645	6.236	5.873	5.549	5.259	4.997
42	8.977	8.262	7.647	7.114	6.648	6.238	5.874	5.550	5.260	4.998
43	8.989	8.270	7.652	7.117	6.650	6.240	5.875	5.551	5.260	4.998
44	8.999	8.276	7.657	7.120	6.652	6.241	5.876	5.552	5.261	4.999
45	9.008	8.282	7.661	7.123	6.654	6.242	5.877	5.552	5.261	4.999
46	9.016	8.288	7.665	7.126	6.656	6.243	5.878	5.553	5.262	4.999
47	9.023	8.293	7.668	7.128	6.657	6.244	5.879	5.553	5.262	4.999
48	9.030	8.297	7.671	7.130	6.658	6.245	5.879	5.554	5.262	5.000
49	9.036	8.301	7.673	7.131	6.659	6.246	5.880	5.554	5.262	5.000
50	9.042	8.304	7.675	7.133	6.660	6.246	5.880	5.554	5.262	5.000

Table 40 Cumulative Present Value Factors - continued

Discount range 21% to 30%.

Year	x	21%	22%	23%	24%	25%	26%	27%	28%	29%	30%
1		0.826	0.820	0.813	0.807	0.800	0.794	0.787	0.781	0.775	0.769
2		1.509	1.492	1.474	1.457	1.440	1.424	1.407	1.392	1.376	1.361
3		2.074	2.042	2.011	1.981	1.952	1.924	1.896	1.869	1.842	1.816
4		2.540	2.494	2.448	2.404	2.362	2.320	2.280	2.241	2.203	2.166
5		2.926	2.864	2.804	2.746	2.689	2.635	2.583	2.532	2.483	2.436
6		3.245	3.167	3.092	3.021	2.951	2.885	2.821	2.759	2.700	2.643
7		3.508	3.416	3.327	3.242	3.161	3.083	3.009	2.937	2.868	2.802
8		3.725	3.619	3.518	3.421	3.329	3.241	3.157	3.076	2.999	2.925
9		3.905	3.786	3.673	3.566	3.463	3.366	3.273	3.184	3.100	3.019
10		4.054	3.923	3.799	3.682	3.571	3.465	3.365	3.269	3.178	3.092
11		4.177	4.036	3.902	3.776	3.656	3.544	3.437	3.335	3.239	3.147
12		4.278	4.128	3.985	3.852	3.725	3.606	3.493	3.387	3.286	3.190
13		4.362	4.203	4.053	3.913	3.780	3.656	3.538	3.427	3.322	3.223
14		4.431	4.265	4.108	3.962	3.824	3.695	3.573	3.459	3.351	3.249
15		4.489	4.315	4.153	4.001	3.859	3.726	3.601	3.484	3.373	3.268
16		4.536	4.357	4.190	4.033	3.887	3.751	3.623	3.503	3.390	3.283
17		4.575	4.391	4.219	4.059	3.910	3.771	3.640	3.518	3.403	3.295
18		4.608	4.419	4.243	4.080	3.928	3.786	3.654	3.530	3.413	3.304
19		4.634	4.442	4.263	4.097	3.942	3.799	3.664	3.539	3.421	3.310
20		4.656	4.460	4.279	4.110	3.954	3.809	3.673	3.546	3.427	3.316
21		4.675	4.476	4.292	4.121	3.963	3.816	3.679	3.552	3.432	3.320
22		4.690	4.488	4.302	4.130	3.970	3.823	3.684	3.556	3.435	3.323
23		4.702	4.499	4.311	4.137	3.976	3.827	3.689	3.559	3.438	3.325
24		4.713	4.507	4.318	4.143	3.981	3.831	3.692	3.562	3.441	3.327
25		4.721	4.514	4.323	4.147	3.985	3.834	3.694	3.564	3.442	3.328
26		4.728	4.520	4.328	4.151	3.988	3.837	3.696	3.566	3.444	3.330
27		4.734	4.525	4.332	4.154	3.990	3.839	3.698	3.567	3.445	3.330
28		4.739	4.528	4.335	4.157	3.992	3.840	3.699	3.568	3.445	3.331
29		4.743	4.531	4.337	4.159	3.994	3.842	3.700	3.569	3.446	3.331
30		4.746	4.534	4.339	4.160	3.995	3.843	3.701	3.570	3.446	3.332
31		4.749	4.536	4.341	4.161	3.996	3.843	3.701	3.570	3.447	3.332
32		4.751	4.538	4.342	4.162	3.997	3.844	3.702	3.570	3.447	3.332
33		4.753	4.539	4.343	4.163	3.997	3.844	3.702	3.571	3.447	3.333
34		4.754	4.540	4.344	4.164	3.998	3.845	3.703	3.571	3.448	3.333
35		4.756	4.541	4.345	4.164	3.998	3.845	3.703	3.571	3.448	3.333
36		4.757	4.542	4.345	4.165	3.998	3.845	3.703	3.571	3.448	3.333
37		4.757	4.543	4.346	4.165	3.999	3.846	3.703	3.571	3.448	3.333
38		4.758	4.543	4.346	4.165	3.999	3.846	3.703	3.571	3.448	3.333
39		4.759	4.544	4.347	4.166	3.999	3.846	3.703	3.572	3.448	3.333
40		4.759	4.544	4.347	4.166	3.999	3.846	3.703	3.572	3.448	3.333
41		4.760	4.544	4.347	4.166	3.999	3.846	3.704	3.572	3.448	3.333
42		4.760	4.545	4.347	4.166	3.999	3.846	3.704	3.572	3.448	3.333
43		4.760	4.545	4.347	4.166	4.000	3.846	3.704	3.572	3.448	3.333
44		4.760	4.545	4.348	4.166	4.000	3.846	3.704	3.572	3.448	3.333
45		4.761	4.545	4.348	4.166	4.000	3.846	3.704	3.572	3.448	3.333
46		4.761	4.545	4.348	4.166	4.000	3.846	3.704	3.572	3.448	3.333
47		4.761	4.545	4.348	4.166	4.000	3.846	3.704	3.572	3.448	3.333
48		4.761	4.545	4.348	4.166	4.000	3.846	3.704	3.572	3.448	3.333
49		4.761	4.545	4.348	4.166	4.000	3.846	3.704	3.572	3.448	3.333
50		4.761	4.545	4.348	4.166	4.000	3.846	3.704	3.572	3.448	3.333

Table 40 Cumulative Present Value Factors – continued

Discount range 31% to 40%.

x Year	31%	32%	33%	34%	35%	36%	37%	38%	39%	40%
1	0.763	0.758	0.752	0.746	0.741	0.735	0.730	0.725	0.719	0.714
2	1.346	1.332	1.317	1.303	1.289	1.276	1.263	1.250	1.237	1.225
3	1.791	1.766	1.742	1.719	1.696	1.674	1.652	1.630	1.609	1.589
4	2.131	2.096	2.062	2.029	1.997	1.966	1.936	1.906	1.877	1.849
5	2.390	2.345	2.302	2.261	2.220	2.181	2.143	2.106	2.070	2.035
6	2.588	2.534	2.483	2.433	2.385	2.339	2.294	2.251	2.209	2.168
7	2.739	2.677	2.619	2.562	2.508	2.455	2.404	2.355	2.308	2.263
8	2.854	2.786	2.721	2.658	2.598	2.540	2.485	2.431	2.380	2.331
9	2.942	2.868	2.798	2.730	2.665	2.603	2.544	2.487	2.432	2.379
10	3.009	2.930	2.855	2.784	2.715	2.649	2.587	2.526	2.469	2.414
11	3.060	2.978	2.899	2.824	2.752	2.683	2.618	2.555	2.496	2.438
12	3.100	3.013	2.931	2.854	2.779	2.708	2.641	2.576	2.515	2.456
13	3.130	3.040	2.956	2.876	2.799	2.727	2.658	2.592	2.529	2.469
14	3.152	3.061	2.974	2.892	2.814	2.740	2.670	2.603	2.538	2.478
15	3.170	3.076	2.988	2.905	2.825	2.750	2.679	2.611	2.546	2.484
16	3.183	3.088	2.999	2.914	2.834	2.757	2.685	2.616	2.551	2.489
17	3.193	3.097	3.006	2.921	2.840	2.763	2.690	2.621	2.554	2.492
18	3.201	3.104	3.012	2.926	2.844	2.767	2.693	2.624	2.557	2.494
19	3.207	3.109	3.017	2.930	2.847	2.770	2.696	2.626	2.559	2.496
20	3.211	3.113	3.020	2.933	2.850	2.772	2.698	2.627	2.560	2.497
21	3.215	3.116	3.023	2.935	2.852	2.773	2.699	2.629	2.561	2.498
22	3.217	3.118	3.024	2.937	2.853	2.775	2.700	2.629	2.562	2.499
23	3.219	3.120	3.026	2.938	2.854	2.775	2.701	2.630	2.563	2.499
24	3.221	3.121	3.027	2.939	2.855	2.776	2.701	2.630	2.563	2.499
25	3.222	3.122	3.028	2.939	2.855	2.776	2.702	2.631	2.563	2.499
26	3.223	3.123	3.028	2.940	2.856	2.777	2.702	2.631	2.564	2.500
27	3.224	3.123	3.029	2.940	2.856	2.777	2.702	2.631	2.564	2.500
28	3.224	3.124	3.029	2.941	2.856	2.777	2.702	2.631	2.564	2.500
29	3.224	3.124	3.029	2.941	2.857	2.777	2.702	2.631	2.564	2.500
30	3.225	3.124	3.030	2.941	2.857	2.777	2.702	2.631	2.564	2.500
31	3.225	3.124	3.030	2.941	2.857	2.777	2.702	2.631	2.564	2.500
32	3.225	3.125	3.030	2.941	2.857	2.778	2.702	2.631	2.564	2.500
33	3.225	3.125	3.030	2.941	2.857	2.778	2.702	2.631	2.564	2.500
34	3.225	3.125	3.030	2.941	2.857	2.778	2.702	2.631	2.564	2.500
35	3.225	3.125	3.030	2.941	2.857	2.778	2.702	2.631	2.564	2.500
36	3.226	3.125	3.030	2.941	2.857	2.778	2.702	2.631	2.564	2.500
37	3.226	3.125	3.030	2.941	2.857	2.778	2.702	2.631	2.564	2.500
38	3.226	3.125	3.030	2.941	2.857	2.778	2.702	2.631	2.564	2.500
39	3.226	3.125	3.030	2.941	2.857	2.778	2.702	2.631	2.564	2.500
40	3.226	3.125	3.030	2.941	2.857	2.778	2.702	2.631	2.564	2.500
41	3.226	3.125	3.030	2.941	2.857	2.778	2.702	2.631	2.564	2.500
42	3.226	3.125	3.030	2.941	2.857	2.778	2.702	2.631	2.564	2.500
43	3.226	3.125	3.030	2.941	2.857	2.778	2.702	2.631	2.564	2.500
44	3.226	3.125	3.030	2.941	2.857	2.778	2.702	2.631	2.564	2.500
45	3.226	3.125	3.030	2.941	2.857	2.778	2.702	2.631	2.564	2.500
46	3.226	3.125	3.030	2.941	2.857	2.778	2.702	2.631	2.564	2.500
47	3.226	3.125	3.030	2.941	2.857	2.778	2.702	2.631	2.564	2.500
48	3.226	3.125	3.030	2.941	2.857	2.778	2.702	2.631	2.564	2.500
49	3.226	3.125	3.030	2.941	2.857	2.778	2.702	2.631	2.564	2.500
50	3.226	3.125	3.030	2.941	2.857	2.778	2.702	2.631	2.564	2.500

Table 41 Capital Recovery Factors

The table gives the equal annual payment to be made for *n* years in the future to repay loan principal and interest with interest at *x*% per year.

For example, to repay £1 borrowed now at 7% in 6 equal annual payments, then the value of each annual payment is £0.2098 or 20.98p. Notice that this is the same thing as the reciprocal of the equivalent present value factor (Table 38).

x Year	1%	2%	3%	4%	5%	6%	7%	8%	9%	10%
1	1.0100	1.0200	1.0300	1.0400	1.0500	1.0600	1.0700	1.0800	1.0900	1.1000
2	0.5075	0.5150	0.5226	0.5302	0.5378	0.5454	0.5531	0.5608	0.5685	0.5762
3	0.3400	0.3468	0.3535	0.3603	0.3672	0.3741	0.3811	0.3880	0.3951	0.4021
4	0.2563	0.2626	0.2690	0.2755	0.2820	0.2886	0.2952	0.3019	0.3087	0.3155
5	0.2060	0.2122	0.2184	0.2246	0.2310	0.2374	0.2439	0.2505	0.2571	0.2638
6	0.1725	0.1785	0.1846	0.1908	0.1970	0.2034	0.2098	0.2163	0.2229	0.2296
7	0.1486	0.1545	0.1605	0.1666	0.1728	0.1791	0.1856	0.1921	0.1987	0.2054
8	0.1307	0.1365	0.1425	0.1485	0.1547	0.1610	0.1675	0.1740	0.1807	0.1874
9	0.1167	0.1225	0.1284	0.1345	0.1407	0.1470	0.1535	0.1601	0.1668	0.1736
10	0.1056	0.1113	0.1172	0.1233	0.1295	0.1359	0.1424	0.1490	0.1558	0.1627
11	0.0965	0.1022	0.1081	0.1141	0.1204	0.1268	0.1334	0.1401	0.1469	0.1540
12	0.0888	0.0946	0.1005	0.1066	0.1128	0.1193	0.1259	0.1327	0.1397	0.1468
13	0.0824	0.0881	0.0940	0.1001	0.1065	0.1130	0.1197	0.1265	0.1336	0.1408
14	0.0769	0.0826	0.0885	0.0947	0.1010	0.1076	0.1143	0.1213	0.1284	0.1357
15	0.0721	0.0778	0.0838	0.0899	0.0963	0.1030	0.1098	0.1168	0.1241	0.1315
16	0.0679	0.0737	0.0796	0.0858	0.0923	0.0990	0.1059	0.1130	0.1203	0.1278
17	0.0643	0.0700	0.0760	0.0822	0.0887	0.0954	0.1024	0.1096	0.1170	0.1247
18	0.0610	0.0667	0.0727	0.0790	0.0855	0.0924	0.0994	0.1067	0.1142	0.1219
19	0.0581	0.0638	0.0698	0.0761	0.0827	0.0896	0.0968	0.1041	0.1117	0.1195
20	0.0554	0.0612	0.0672	0.0736	0.0802	0.0872	0.0944	0.1019	0.1095	0.1175
21	0.0530	0.0588	0.0649	0.0713	0.0780	0.0850	0.0923	0.0998	0.1076	0.1156
22	0.0509	0.0566	0.0627	0.0692	0.0760	0.0830	0.0904	0.0980	0.1059	0.1140
23	0.0489	0.0547	0.0608	0.0673	0.0741	0.0813	0.0887	0.0964	0.1044	0.1126
24	0.0471	0.0529	0.0590	0.0656	0.0725	0.0797	0.0872	0.0950	0.1030	0.1113
25	0.0454	0.0512	0.0574	0.0640	0.0710	0.0782	0.0858	0.0937	0.1018	0.1102
26	0.0439	0.0497	0.0559	0.0626	0.0696	0.0769	0.0846	0.0925	0.1007	0.1092
27	0.0424	0.0483	0.0546	0.0612	0.0683	0.0757	0.0834	0.0914	0.0997	0.1083
28	0.0411	0.0470	0.0533	0.0600	0.0671	0.0746	0.0824	0.0905	0.0989	0.1075
29	0.0399	0.0458	0.0521	0.0589	0.0660	0.0736	0.0814	0.0896	0.0981	0.1067
30	0.0387	0.0446	0.0510	0.0578	0.0651	0.0726	0.0806	0.0888	0.0973	0.1061
31	0.0377	0.0436	0.0500	0.0569	0.0641	0.0718	0.0798	0.0881	0.0967	0.1055
32	0.0367	0.0426	0.0490	0.0559	0.0633	0.0710	0.0791	0.0875	0.0961	0.1050
33	0.0357	0.0417	0.0482	0.0551	0.0625	0.0703	0.0784	0.0869	0.0956	0.1045
34	0.0348	0.0408	0.0473	0.0543	0.0618	0.0696	0.0778	0.0863	0.0951	0.1041
35	0.0340	0.0400	0.0465	0.0536	0.0611	0.0690	0.0772	0.0858	0.0946	0.1037
36	0.0332	0.0392	0.0458	0.0529	0.0604	0.0684	0.0767	0.0853	0.0942	0.1033
37	0.0325	0.0385	0.0451	0.0522	0.0598	0.0679	0.0762	0.0849	0.0939	0.1030
38	0.0318	0.0378	0.0445	0.0516	0.0593	0.0674	0.0758	0.0845	0.0935	0.1027
39	0.0311	0.0372	0.0438	0.0511	0.0588	0.0669	0.0754	0.0842	0.0932	0.1025
40	0.0305	0.0366	0.0433	0.0505	0.0583	0.0665	0.0750	0.0839	0.0930	0.1023
41	0.0299	0.0360	0.0427	0.0500	0.0578	0.0661	0.0747	0.0836	0.0927	0.1020
42	0.0293	0.0354	0.0422	0.0495	0.0574	0.0657	0.0743	0.0833	0.0925	0.1019
43	0.0287	0.0349	0.0417	0.0491	0.0570	0.0653	0.0740	0.0830	0.0923	0.1017
44	0.0282	0.0344	0.0412	0.0487	0.0566	0.0650	0.0738	0.0828	0.0921	0.1015
45	0.0277	0.0339	0.0408	0.0483	0.0563	0.0647	0.0735	0.0826	0.0919	0.1014
46	0.0272	0.0335	0.0404	0.0479	0.0559	0.0644	0.0733	0.0824	0.0917	0.1013
47	0.0268	0.0330	0.0400	0.0475	0.0556	0.0641	0.0730	0.0822	0.0916	0.1011
48	0.0263	0.0326	0.0396	0.0472	0.0553	0.0639	0.0728	0.0820	0.0915	0.1010
49	0.0259	0.0322	0.0392	0.0469	0.0550	0.0637	0.0726	0.0819	0.0913	0.1009
50	0.0255	0.0318	0.0389	0.0466	0.0548	0.0634	0.0725	0.0817	0.0912	0.1009

Table 41 Capital Recovery Factors – continued

Interest range 11% to 20%

x Year	11%	12%	13%	14%	15%	16%	17%	18%	19%	20%
1	1.1100	1.1200	1.1300	1.1400	1.1500	1.1600	1.1700	1.1800	1.1900	1.2000
2	0.5839	0.5917	0.5995	0.6073	0.6151	0.6230	0.6308	0.6387	0.6466	0.6545
3	0.4092	0.4163	0.4235	0.4307	0.4380	0.4453	0.4526	0.4599	0.4673	0.4747
4	0.3223	0.3292	0.3362	0.3432	0.3503	0.3574	0.3645	0.3717	0.3790	0.3863
5	0.2706	0.2774	0.2843	0.2913	0.2983	0.3054	0.3126	0.3198	0.3271	0.3344
6	0.2364	0.2432	0.2502	0.2572	0.2642	0.2714	0.2786	0.2859	0.2933	0.3007
7	0.2122	0.2191	0.2261	0.2332	0.2404	0.2476	0.2549	0.2624	0.2699	0.2774
8	0.1943	0.2013	0.2084	0.2156	0.2229	0.2302	0.2377	0.2452	0.2529	0.2606
9	0.1806	0.1877	0.1949	0.2022	0.2096	0.2171	0.2247	0.2324	0.2402	0.2481
10	0.1698	0.1770	0.1843	0.1917	0.1993	0.2069	0.2147	0.2225	0.2305	0.2385
11	0.1611	0.1684	0.1758	0.1834	0.1911	0.1989	0.2068	0.2148	0.2229	0.2311
12	0.1540	0.1614	0.1690	0.1767	0.1845	0.1924	0.2005	0.2086	0.2169	0.2253
13	0.1482	0.1557	0.1634	0.1712	0.1791	0.1872	0.1954	0.2037	0.2121	0.2206
14	0.1432	0.1509	0.1587	0.1666	0.1747	0.1829	0.1912	0.1997	0.2082	0.2169
15	0.1391	0.1468	0.1547	0.1628	0.1710	0.1794	0.1878	0.1964	0.2051	0.2139
16	0.1355	0.1434	0.1514	0.1596	0.1679	0.1764	0.1850	0.1937	0.2025	0.2114
17	0.1325	0.1405	0.1486	0.1569	0.1654	0.1740	0.1827	0.1915	0.2004	0.2094
18	0.1298	0.1379	0.1462	0.1546	0.1632	0.1719	0.1807	0.1896	0.1987	0.2078
19	0.1276	0.1358	0.1441	0.1527	0.1613	0.1701	0.1791	0.1881	0.1972	0.2065
20	0.1256	0.1339	0.1424	0.1510	0.1598	0.1687	0.1777	0.1868	0.1960	0.2054
21	0.1238	0.1322	0.1408	0.1495	0.1584	0.1674	0.1765	0.1857	0.1951	0.2044
22	0.1223	0.1308	0.1395	0.1483	0.1573	0.1664	0.1756	0.1848	0.1942	0.2037
23	0.1210	0.1296	0.1383	0.1472	0.1563	0.1654	0.1747	0.1841	0.1935	0.2031
24	0.1198	0.1285	0.1373	0.1463	0.1554	0.1647	0.1740	0.1835	0.1930	0.2025
25	0.1187	0.1275	0.1364	0.1455	0.1547	0.1640	0.1734	0.1829	0.1925	0.2021
26	0.1178	0.1267	0.1357	0.1448	0.1541	0.1634	0.1729	0.1825	0.1921	0.2018
27	0.1170	0.1259	0.1350	0.1442	0.1535	0.1630	0.1725	0.1821	0.1917	0.2015
28	0.1163	0.1252	0.1344	0.1437	0.1531	0.1625	0.1721	0.1818	0.1915	0.2012
29	0.1156	0.1247	0.1339	0.1432	0.1527	0.1622	0.1718	0.1815	0.1912	0.2010
30	0.1150	0.1241	0.1334	0.1428	0.1523	0.1619	0.1715	0.1813	0.1910	0.2008
31	0.1145	0.1237	0.1330	0.1425	0.1520	0.1616	0.1713	0.1811	0.1909	0.2007
32	0.1140	0.1233	0.1327	0.1421	0.1517	0.1614	0.1711	0.1809	0.1907	0.2006
33	0.1136	0.1229	0.1323	0.1419	0.1515	0.1612	0.1710	0.1808	0.1906	0.2005
34	0.1133	0.1226	0.1321	0.1416	0.1513	0.1610	0.1708	0.1806	0.1905	0.2004
35	0.1129	0.1223	0.1318	0.1414	0.1511	0.1609	0.1707	0.1806	0.1904	0.2003
36	0.1126	0.1221	0.1316	0.1413	0.1510	0.1608	0.1706	0.1805	0.1904	0.2003
37	0.1124	0.1218	0.1314	0.1411	0.1509	0.1607	0.1705	0.1804	0.1903	0.2002
38	0.1121	0.1216	0.1313	0.1410	0.1507	0.1606	0.1704	0.1803	0.1903	0.2002
39	0.1119	0.1215	0.1311	0.1409	0.1506	0.1605	0.1704	0.1803	0.1902	0.2002
40	0.1117	0.1213	0.1310	0.1407	0.1506	0.1604	0.1703	0.1802	0.1902	0.2001
41	0.1115	0.1212	0.1309	0.1407	0.1505	0.1604	0.1703	0.1802	0.1902	0.2001
42	0.1114	0.1210	0.1308	0.1406	0.1504	0.1603	0.1702	0.1802	0.1901	0.2001
43	0.1113	0.1209	0.1307	0.1405	0.1504	0.1603	0.1702	0.1801	0.1901	0.2001
44	0.1111	0.1208	0.1306	0.1404	0.1503	0.1602	0.1702	0.1801	0.1901	0.2001
45	0.1110	0.1207	0.1305	0.1404	0.1503	0.1602	0.1701	0.1801	0.1901	0.2001
46	0.1109	0.1207	0.1305	0.1403	0.1502	0.1602	0.1701	0.1801	0.1901	0.2000
47	0.1108	0.1206	0.1304	0.1403	0.1502	0.1601	0.1701	0.1801	0.1901	0.2000
48	0.1107	0.1205	0.1304	0.1403	0.1502	0.1601	0.1701	0.1801	0.1900	0.2000
49	0.1107	0.1205	0.1303	0.1402	0.1502	0.1601	0.1701	0.1801	0.1900	0.2000
50	0.1106	0.1204	0.1303	0.1402	0.1501	0.1601	0.1701	0.1800	0.1900	0.2000

Table 41 Capital Recovery Factors – continued

Interest range 21% to 30%

x Year	21%	22%	23%	24%	25%	26%	27%	28%	29%	30%
1	1.2100	1.2200	1.2300	1.2400	1.2500	1.2600	1.2700	1.2800	1.2900	1.3000
2	0.6625	0.6705	0.6784	0.6864	0.6944	0.7025	0.7105	0.7186	0.7267	0.7348
3	0.4822	0.4897	0.4972	0.5047	0.5123	0.5199	0.5275	0.5352	0.5429	0.5506
4	0.3936	0.4010	0.4085	0.4159	0.4234	0.4310	0.4386	0.4462	0.4539	0.4616
5	0.3418	0.3492	0.3567	0.3642	0.3718	0.3795	0.3872	0.3949	0.4027	0.4106
6	0.3082	0.3158	0.3234	0.3311	0.3388	0.3466	0.3545	0.3624	0.3704	0.3784
7	0.2851	0.2928	0.3006	0.3084	0.3163	0.3243	0.3324	0.3405	0.3486	0.3569
8	0.2684	0.2763	0.2843	0.2923	0.3004	0.3086	0.3168	0.3251	0.3335	0.3419
9	0.2561	0.2641	0.2722	0.2805	0.2888	0.2971	0.3056	0.3140	0.3226	0.3312
10	0.2467	0.2549	0.2632	0.2716	0.2801	0.2886	0.2972	0.3059	0.3147	0.3235
11	0.2394	0.2478	0.2563	0.2649	0.2735	0.2822	0.2910	0.2998	0.3088	0.3177
12	0.2337	0.2423	0.2509	0.2596	0.2684	0.2773	0.2863	0.2953	0.3043	0.3135
13	0.2292	0.2379	0.2467	0.2556	0.2645	0.2736	0.2826	0.2918	0.3010	0.3102
14	0.2256	0.2345	0.2434	0.2524	0.2615	0.2706	0.2799	0.2891	0.2984	0.3078
15	0.2228	0.2317	0.2408	0.2499	0.2591	0.2684	0.2777	0.2871	0.2965	0.3060
16	0.2204	0.2295	0.2387	0.2479	0.2572	0.2666	0.2760	0.2855	0.2950	0.3046
17	0.2186	0.2278	0.2370	0.2464	0.2558	0.2652	0.2747	0.2843	0.2939	0.3035
18	0.2170	0.2263	0.2357	0.2451	0.2546	0.2641	0.2737	0.2833	0.2930	0.3027
19	0.2158	0.2251	0.2346	0.2441	0.2537	0.2633	0.2729	0.2826	0.2923	0.3021
20	0.2147	0.2242	0.2337	0.2433	0.2529	0.2626	0.2723	0.2820	0.2918	0.3016
21	0.2139	0.2234	0.2330	0.2426	0.2523	0.2620	0.2718	0.2816	0.2914	0.3012
22	0.2132	0.2228	0.2324	0.2421	0.2519	0.2616	0.2714	0.2812	0.2911	0.3009
23	0.2127	0.2223	0.2320	0.2417	0.2515	0.2613	0.2711	0.2810	0.2908	0.3007
24	0.2122	0.2219	0.2316	0.2414	0.2512	0.2610	0.2709	0.2808	0.2906	0.3006
25	0.2118	0.2215	0.2313	0.2411	0.2509	0.2608	0.2707	0.2806	0.2905	0.3004
26	0.2115	0.2213	0.2311	0.2409	0.2508	0.2606	0.2705	0.2805	0.2904	0.3003
27	0.2112	0.2210	0.2309	0.2407	0.2506	0.2605	0.2704	0.2804	0.2903	0.3003
28	0.2110	0.2208	0.2307	0.2406	0.2505	0.2604	0.2703	0.2803	0.2902	0.3002
29	0.2108	0.2207	0.2306	0.2405	0.2504	0.2603	0.2703	0.2802	0.2902	0.3001
30	0.2107	0.2206	0.2305	0.2404	0.2503	0.2603	0.2702	0.2802	0.2901	0.3001
31	0.2106	0.2205	0.2304	0.2403	0.2502	0.2602	0.2702	0.2801	0.2901	0.3001
32	0.2105	0.2204	0.2303	0.2402	0.2502	0.2602	0.2701	0.2801	0.2901	0.3001
33	0.2104	0.2203	0.2302	0.2402	0.2502	0.2601	0.2701	0.2801	0.2901	0.3001
34	0.2103	0.2203	0.2302	0.2402	0.2501	0.2601	0.2701	0.2801	0.2901	0.3000
35	0.2103	0.2202	0.2302	0.2401	0.2501	0.2601	0.2701	0.2800	0.2900	0.3000
36	0.2102	0.2202	0.2301	0.2401	0.2501	0.2601	0.2700	0.2800	0.2900	0.3000
37	0.2102	0.2201	0.2301	0.2401	0.2501	0.2601	0.2700	0.2800	0.2900	0.3000
38	0.2102	0.2201	0.2301	0.2401	0.2501	0.2600	0.2700	0.2800	0.2900	0.3000
39	0.2101	0.2201	0.2301	0.2401	0.2500	0.2600	0.2700	0.2800	0.2900	0.3000
40	0.2101	0.2201	0.2301	0.2400	0.2500	0.2600	0.2700	0.2800	0.2900	0.3000
41	0.2101	0.2201	0.2300	0.2400	0.2500	0.2600	0.2700	0.2800	0.2900	0.3000
42	0.2101	0.2201	0.2300	0.2400	0.2500	0.2600	0.2700	0.2800	0.2900	0.3000
43	0.2101	0.2200	0.2300	0.2400	0.2500	0.2600	0.2700	0.2800	0.2900	0.3000
44	0.2100	0.2200	0.2300	0.2400	0.2500	0.2600	0.2700	0.2800	0.2900	0.3000
45	0.2100	0.2200	0.2300	0.2400	0.2500	0.2600	0.2700	0.2800	0.2900	0.3000
46	0.2100	0.2200	0.2300	0.2400	0.2500	0.2600	0.2700	0.2800	0.2900	0.3000
47	0.2100	0.2200	0.2300	0.2400	0.2500	0.2600	0.2700	0.2800	0.2900	0.3000
48	0.2100	0.2200	0.2300	0.2400	0.2500	0.2600	0.2700	0.2800	0.2900	0.3000
49	0.2100	0.2200	0.2300	0.2400	0.2500	0.2600	0.2700	0.2800	0.2900	0.3000
50	0.2100	0.2200	0.2300	0.2400	0.2500	0.2600	0.2700	0.2800	0.2900	0.3000

Table 41 Capital Recovery Factors – continued

Interest range 31% to 40%

x Year	31%	32%	33%	34%	35%	36%	37%	38%	39%	40%
1	1.3100	1.3200	1.3300	1.3400	1.3500	1.3600	1.3700	1.3800	1.3900	1.4000
2	0.7429	0.7510	0.7592	0.7674	0.7755	0.7837	0.7919	0.8002	0.8084	0.8167
3	0.5584	0.5662	0.5740	0.5818	0.5897	0.5976	0.6055	0.6134	0.6214	0.6294
4	0.4694	0.4772	0.4850	0.4929	0.5008	0.5087	0.5167	0.5247	0.5327	0.5408
5	0.4185	0.4264	0.4344	0.4424	0.4505	0.4586	0.4667	0.4749	0.4831	0.4914
6	0.3865	0.3946	0.4028	0.4110	0.4193	0.4276	0.4359	0.4443	0.4528	0.4613
7	0.3652	0.3735	0.3819	0.3903	0.3988	0.4073	0.4159	0.4245	0.4332	0.4419
8	0.3504	0.3589	0.3675	0.3762	0.3849	0.3936	0.4024	0.4113	0.4201	0.4291
9	0.3399	0.3487	0.3575	0.3663	0.3752	0.3841	0.3931	0.4022	0.4112	0.4203
10	0.3323	0.3412	0.3502	0.3592	0.3683	0.3774	0.3866	0.3958	0.4050	0.4143
11	0.3268	0.3358	0.3450	0.3542	0.3634	0.3727	0.3820	0.3913	0.4007	0.4101
12	0.3226	0.3319	0.3411	0.3505	0.3598	0.3692	0.3787	0.3881	0.3976	0.4072
13	0.3196	0.3289	0.3383	0.3477	0.3572	0.3667	0.3763	0.3859	0.3955	0.4051
14	0.3172	0.3267	0.3362	0.3457	0.3553	0.3649	0.3746	0.3842	0.3939	0.4036
15	0.3155	0.3251	0.3346	0.3443	0.3539	0.3636	0.3733	0.3831	0.3928	0.4026
16	0.3142	0.3238	0.3335	0.3432	0.3529	0.3626	0.3724	0.3822	0.3920	0.4018
17	0.3132	0.3229	0.3326	0.3424	0.3521	0.3619	0.3718	0.3816	0.3915	0.4013
18	0.3124	0.3222	0.3320	0.3418	0.3516	0.3614	0.3713	0.3812	0.3910	0.4009
19	0.3118	0.3216	0.3315	0.3413	0.3512	0.3610	0.3709	0.3808	0.3907	0.4007
20	0.3114	0.3212	0.3311	0.3410	0.3509	0.3608	0.3707	0.3806	0.3905	0.4005
21	0.3111	0.3209	0.3308	0.3407	0.3506	0.3606	0.3705	0.3804	0.3904	0.4003
22	0.3108	0.3207	0.3306	0.3405	0.3505	0.3604	0.3704	0.3803	0.3903	0.4002
23	0.3106	0.3205	0.3305	0.3404	0.3504	0.3603	0.3703	0.3802	0.3902	0.4002
24	0.3105	0.3204	0.3304	0.3403	0.3503	0.3602	0.3702	0.3802	0.3901	0.4001
25	0.3104	0.3203	0.3303	0.3402	0.3502	0.3602	0.3701	0.3801	0.3901	0.4001
26	0.3103	0.3202	0.3302	0.3402	0.3501	0.3601	0.3701	0.3801	0.3901	0.4001
27	0.3102	0.3202	0.3301	0.3401	0.3501	0.3601	0.3701	0.3801	0.3901	0.4000
28	0.3102	0.3201	0.3301	0.3401	0.3501	0.3601	0.3701	0.3800	0.3900	0.4000
29	0.3101	0.3201	0.3301	0.3401	0.3501	0.3600	0.3700	0.3800	0.3900	0.4000
30	0.3101	0.3201	0.3301	0.3401	0.3500	0.3600	0.3700	0.3800	0.3900	0.4000
31	0.3101	0.3201	0.3300	0.3400	0.3500	0.3600	0.3700	0.3800	0.3900	0.4000
32	0.3101	0.3200	0.3300	0.3400	0.3500	0.3600	0.3700	0.3800	0.3900	0.4000
33	0.3100	0.3200	0.3300	0.3400	0.3500	0.3600	0.3700	0.3800	0.3900	0.4000
34	0.3100	0.3200	0.3300	0.3400	0.3500	0.3600	0.3700	0.3800	0.3900	0.4000
35	0.3100	0.3200	0.3300	0.3400	0.3500	0.3600	0.3700	0.3800	0.3900	0.4000
36	0.3100	0.3200	0.3300	0.3400	0.3500	0.3600	0.3700	0.3800	0.3900	0.4000
37	0.3100	0.3200	0.3300	0.3400	0.3500	0.3600	0.3700	0.3800	0.3900	0.4000
38	0.3100	0.3200	0.3300	0.3400	0.3500	0.3600	0.3700	0.3800	0.3900	0.4000
39	0.3100	0.3200	0.3300	0.3400	0.3500	0.3600	0.3700	0.3800	0.3900	0.4000
40	0.3100	0.3200	0.3300	0.3400	0.3500	0.3600	0.3700	0.3800	0.3900	0.4000
41	0.3100	0.3200	0.3300	0.3400	0.3500	0.3600	0.3700	0.3800	0.3900	0.4000
42	0.3100	0.3200	0.3300	0.3400	0.3500	0.3600	0.3700	0.3800	0.3900	0.4000
43	0.3100	0.3200	0.3300	0.3400	0.3500	0.3600	0.3700	0.3800	0.3900	0.4000
44	0.3100	0.3200	0.3300	0.3400	0.3500	0.3600	0.3700	0.3800	0.3900	0.4000
45	0.3100	0.3200	0.3300	0.3400	0.3500	0.3600	0.3700	0.3800	0.3900	0.4000
46	0.3100	0.3200	0.3300	0.3400	0.3500	0.3600	0.3700	0.3800	0.3900	0.4000
47	0.3100	0.3200	0.3300	0.3400	0.3500	0.3600	0.3700	0.3800	0.3900	0.4000
48	0.3100	0.3200	0.3300	0.3400	0.3500	0.3600	0.3700	0.3800	0.3900	0.4000
49	0.3100	0.3200	0.3300	0.3400	0.3500	0.3600	0.3700	0.3800	0.3900	0.4000
50	0.3100	0.3200	0.3300	0.3400	0.3500	0.3600	0.3700	0.3800	0.3900	0.4000

Examples of the Use of Tables 11 to 16

Using Table 11 – Tukey's Wholly Significant Difference (5% Level)

When a one-way analysis of various results in a significantly high ratio of mean squares, the need is to determine which of the group means may be responsible.

If there were only two means, $\bar{x}_1$ and $\bar{x}_2$, involved, then since

$$t = \frac{(\bar{x}_1 - \bar{x}_2)}{s\sqrt{\left(\frac{1}{n} + \frac{1}{n}\right)}} = \frac{(\bar{x}_1 - \bar{x}_2)}{s\frac{\sqrt{2}}{\sqrt{n}}} = \frac{(\bar{x}_1 - \bar{x}_2)}{s(\bar{x})\sqrt{2}}$$

where s^2 is the residual variance, n is the (same) number of readings in each mean, and $s(\bar{x})$ is the estimated standard error of the mean of n readings.

$|\bar{x}_1 - \bar{x}_2|$ is significant if it is greater than $t\sqrt{2}.s(\bar{x})$

$t\sqrt{2}.s(\bar{x})$ defines a Least Significant Difference (LSD) between means at any chosen significance level. In repeated sampling, at the 5% level, $|\bar{x}_1 - \bar{x}_2|$ would be greater than the LSD one time in 20 in the long run when there was no actual difference between the two population means.

With k groups, there are $\dfrac{k(k-1)}{2}$ paired comparisons and there is a considerable chance (0.54 for $k = 6$) that at least one of them will exceed the LSD even when all k population means are identical.

Tukey has proposed the term 'Wholly Significant Difference' (WSD) such that if the difference $|\bar{x}_i - \bar{x}_j|$ between any pair of sample means exceeds this WSD, then the given population means are assumed to be different at the relevant significant level.

Table 11 contains values of WSD/$s(\bar{x})$ tabulated for k groups and pooled error variance degrees of freedom, v.

Example

Group	1	2	3	4	5	6
	29	35	40	30	25	41
	23	32	33	32	28	32
	22	31	37	34	27	35
	25	29	35	34	28	32
	27	32	37	30	25	32
Totals	126	159	182	160	133	172
Means	25.2	31.8	36.4	32.0	26.6	34.4

The analysis of variance table for the data is:

Source of Variation	Sums of Squares	Degrees of Freedom	Mean Squares
Between groups	476.6	5	95.3
Within groups	165.2	24	6.9
Total	641.9	29	

The observed ratio of mean squares (95.3/6.9 = 13.8) exceeds $F_{.001,5,24}$ (5.98 from Table 9) and there is thus strong evidence that the group means are not all equal.

We can use Tukey's WSD to assess which groups are most likely to be responsible for this result.

We have $k = 6$, $n = 5$ and $s = \sqrt{6.9}$ with $v = 24$.

The standard error of a group sample mean $s(\bar{x}) = \sqrt{\dfrac{6.9}{5}} = \underline{1.17}$

and from Table 11, WSD/$s(\bar{x})$ = 4.40 by linear interpolation.

For a statistically significant difference, two sample means must differ by at least the WSD, which for the present data is $4.40 \times s(\bar{x}) = 4.40 \times 1.17 = \underline{5.1}$.

The next table shows, in a systematic arrangement, all paired comparisons between the group means; differences greater than the WSD are marked with an asterisk.

		Group					
		1	5	2	4	6	3
Group	Mean	25.2	26.6	31.8	32.0	34.4	36.4
1	25.2	–	1.4	6.6*	6.8*	9.2*	11.2*
5	26.6		–	5.2*	5.4*	7.8*	9.8*
2	31.8			–	0.2	2.6	4.6
4	32.0				–	2.4	4.4
6	34.4					–	2.0
3	36.4						–

Thus the evidence is that the six populations split into two sets; 1 and 5 are much alike as are 2, 4, 6 and 3 among themselves, the latter having a higher mean than 1 and 5.

Note that the Least Significant Difference of 3.43 based on the incorrect use of t suggests that the mean of populations 2 and 4 differ from 3 but not from 6 and that 6 does not differ from 3, a somewhat inconsistent result.

Remember also that, although the difference in means between the two sets is *statistically* significant, the actual size of the mean difference (estimated as about $33.6 - 25.9 = 7.7$) may be of no *practical* importance whatsoever.

Using Table 12 – Spearman's Rank Correlation Coefficient

Each of two interviewers makes an independent subjective assessment of the suitability of 8 candidates for a given job. Their orders of preference (ranks) are shown in the table. Is there any measure of agreement between the assessors' judgement?

Candidate	Assessor X	Assessor Y	Difference in Ranks $d = X - Y$	d^2
A	4	4	0	0
B	8	7	1	1
C	1	2	–1	1
D	6	8	–2	4
E	2	3	–1	1
F	3	1	2	4
G	5	6	–1	1
H	7	5	2	4
				$\Sigma d^2 = 16$

$$r_S = 1 - \frac{6\Sigma d^2}{n(n^2-1)} = 1 - \frac{6 \times 16}{8 \times 63} = 1 - \frac{4}{21} = \frac{17}{21} = \underline{0.81}$$

Assuming that there is no correlation (agreement) between the opinions of the assessors, Table 12 shows that the 5% point of r_s for $n = 8$ is 0.643 and the 1% point is 0.833.

The observed value of r_s is therefore significant at the 5% level though not quite at the 1% level (one-sided). There is thus a tendency for the interviewers to make similar assessments of job candidates. Note that, as with most distributions based on combinatorial properties of numbers, the probability of r_s being equal to or greater than 0.643 is *at most* 5% rather than *exactly* the 5% p-value specified in the table.

Approximate method

Although 8 pairs of ranks is not large, it is instructive to use Student's t approximation. The observed value of r_s corresponds to:

$$t \approx r_s \sqrt{\frac{n-2}{1-r_s^2}} = 0.81 * \sqrt{\frac{6}{1-0.81^2}} = 3.38$$

Reference to Table 7 shows that this value of t for 6 degrees of freedom is just significant at the 1% level ($t_{.01,6} = 3.14$).

The exact test for r_s showed that it was not quite significant at the 1% level. These results though not exactly the same are comparable (there is nothing special about a probability of 1%). The approximation is certainly good enough for most practical purposes when n is greater than 10.

Using Table 13 – Kendall's Rank Correlation Coefficient

Example

The data used to illustrate the calculation of Spearman's Coefficient can be used again here. First of all it is necessary to rearrange the ranks of one of the assessors, say X, in ascending order. The score S can then be found for each of the resulting pairs of ranks for the second assessor.

	Candidate							
	C	E	F	A	G	D	H	B
Assessor X	1	2	3	4	5	6	7	8
Assessor Y	2	3	1	4	6	8	5	7

Paired Ranks of Y	Score	Paired Ranks of Y	Score	Paired Ranks of Y	Score
$2 \to 3$	+1	$1 \to 4$	+1	$8 \to 5$	−1
$2 \to 1$	−1	$1 \to 6$	+1	$8 \to 7$	−1
$2 \to 4$	+1	$1 \to 8$	+1		
$2 \to 6$	+1	$1 \to 5$	+1	$5 \to 7$	+1
$2 \to 8$	+1	$1 \to 7$	+1		
$2 \to 5$	+1				
$2 \to 7$	+1	$4 \to 6$	+1		
		$4 \to 8$	+1		
$3 \to 1$	−1	$4 \to 5$	+1	$S = +23 - 5 = +18$	
$3 \to 4$	+1	$4 \to 7$	+1	For $n = 8$ pairs of ranks,	
$3 \to 6$	+1			the probability of a	
$3 \to 8$	+1	$6 \to 8$	+1	score of +18 or more	
$3 \to 5$	+1	$6 \to 5$	−1	assuming no correlation	
$3 \to 7$	+1	$6 \to 7$	+1	of ranks, is +0.016	

The result, in agreement with Spearman's test (which is in fact identical in power to Kendall's test), is that the observed score of +18 $\left(r_k = \dfrac{+18}{8 \times 7} \times 2 = 0.64 \right)$ is not quite significant at the 1% level (one-sided alternative). Most observers would no doubt take this result as reasonable evidence of general agreement between the assessors (though their assessments need not correspond to the candidates' actual suitability for the job!).

Approximate method

Although n should preferably be greater than 10, the approximate normality of r_k can be used for this example in order to illustrate the method.

$$Z = S \sqrt{\frac{18}{n(n-1)(2n+5)}}$$ has approximately a unit normal distribution.

Thus: $$Z = +18 \sqrt{\frac{18}{8 \times 7 \times 21}} = 2.23$$

From Table 3, the probability of exceeding $Z = 2.23$ (one-sided test) is 0.013. Note that this is very little different from the exact probability and would lead to the same conclusion about the assessors' similarity.

Identifying Extreme Readings – Tables 14 and 15

In any small group of observations, there may be some doubt as to whether a single outlying reading belongs to the same population as the remainder of the sample.

Although data should not generally be excluded from a sample without good reason, the tables given here provide a means of assessing whether an outlying reading *could* be a 'rogue' observation.

Two tables are given. Table 14 relates to Nair's test which is based on the 'studentised' extreme deviate from the sample mean. Use of this test requires an independent estimate of the standard deviation of the population (assumed normal) to be available external to the sample. For further information, refer to Nair, K.R. (1952), *Biometrika*, 39:190.

Table 15 gives percentage points of the rank difference ratio, the statistic used in what is known as Dixon's test. In this case no independent estimate of the population standard deviation is necessary. For further information, refer to Dixon, W.J. (1951), *Ann. Math. Stat.*, 22:68.

Example: Using Table 14 – Nair's Test

An estimate of the standard deviation of a normal distribution is available based on 25 observations and is equal to 0.3 units. A sample of size 5 subsequently taken from the population gave the following readings:

12.1,   12.4,   12.0,   11.2,   11.8

Can either the lowest or the highest reading be classified as being extreme and therefore probably not generated under the same conditions as the remainder of the sample?

The sample mean, $\bar{x}$, is 11.9 and the lowest reading of 11.2 deviates from this by more than does the highest reading of 12.4.

$$\frac{\left|x_{(1)} - \bar{x}\right|}{s} = \frac{\left|11.2 - 11.9\right|}{0.3} = 2.33$$

Table 14 shows that, for $v = 24$ and $n = 5$, the upper 5% and 1% points are 2.23 and 2.85 respectively. There is thus some suggestion that the observation 11.2 is a low extreme reading (significant at 5% though not at 1%).

Example: Using Table 15 – Dixon's Test

In the following random sample of 9 observations, can the extreme value (6.2) be regarded as coming from the same population as the remainder of the sample?

Random observations x_i :   8.1, 8.3, 7.9, 8.4, 7.9, 8.1, 6.2, 8.2, 8.6
Ranked observations $x_{(i)}$:   6.2, 7.9, 7.9, 8.1, 8.1, 8.2, 8,3, 8.4, 8.6

Since $x_{(1)}$ is the extreme value of interest and $n = 9$, we use the rank difference ratio of

$$\frac{x_{(3)} - x_{(1)}}{x_{(n-1)} - x_{(1)}} = \frac{7.9 - 6.2}{8.4 - 6.2} = \frac{1.7}{2.2} = 0.773$$

Table 15 shows that the 1% point of this ratio is 0.776 and so the observed ratio is significantly high at about the 1% level. There is thus quite good evidence for identifying the observation of 6.2 as being excessively low and therefore worthy of careful scrutiny.

Using Table 16 – The One-Sample Kolmogorov-Smirnov Distribution

This distribution provides a test of goodness of fit.

Example

A six-sided die is rolled 10 times giving scores of:

2, 6, 6, 3, 4, 4, 2, 5, 3, 4.

First, calculate the observed cumulative relative frequency of each possible die score from 1 to 6. Take the difference between each observed relative frequency and the corresponding theoretical cumulative probability given by the fitted model.

Here we would expect the scores to be a random sample from a uniform, that is, rectangular, distribution of integer values between 1 and 6 inclusive. The table below shows the calculations:

Score	1	2	3	4	5	6
Observed cumulative probability	0.00	0.20	0.40	0.70	0.80	1.00
Theoretical cumulative probability	0.17	0.33	0.50	0.67	0.83	1.00
Absolute difference D	0.17	0.13	0.10	0.03	0.03	0.00

The maximum observed absolute difference, D, is 0.17.

From Table 16 for $n = 10$ observations, the 20% point of D is 0.322 and the observed value of D (= 0.17) does not exceed this. Thus there is no evidence to reject our implicit test hypothesis which is that the observed scores are consistent with fair throws of a perfect die.

Some Useful Formulae

Description	Symbol	Definition	Hand Computational Form
Sample Statistics Sample (Arithmetic) Mean or Average	$\bar{x}$	Total of the observed values in the sample divided by the number of observations.	$\dfrac{\sum\limits_{i=1}^{n} x_i}{n}$ or $\dfrac{\sum\limits_{i=1}^{k} f_i x_i}{\sum\limits_{i=1}^{k} f_i}$
Sample Range	R	The difference between the largest and smallest values in the sample.	
Sample Variance	$\hat{s}^2$	$\dfrac{1}{n}\sum\limits_{i=1}^{n}(x_i - \bar{x})^2$ or $\dfrac{1}{\sum\limits_{i=1}^{k} f_i}\sum\limits_{i=1}^{k} f_i(x_i - \bar{x})^2$	$\dfrac{\sum x_i^2 - \dfrac{(\sum x_i)^2}{n}}{n}$ or $\dfrac{\sum f_i x_i^2 - \dfrac{(\sum f_i x_i)^2}{\sum f_i}}{\sum f_i}$
Sample Standard Deviation	$\hat{s}$	Positive square root of sample variance.	
Unbiased estimator of population mean	$\overline{\mu}$	$\bar{x}$	
Unbiased estimate of population variance (population mean, μ, not known)	s^2	$\dfrac{n\hat{s}^2}{n-1} = \dfrac{1}{n-1}\sum\limits_{i=1}^{n}(x_i - \bar{x})^2$	$\dfrac{\sum x_i^2 - \dfrac{(\sum x_i)^2}{n}}{n-1}$
Pooled estimate of two population variances extended to any number of sample variance estimates		$\dfrac{(n_1 - 1)s_1^2 + (n_2 - 1)s_2^2}{n_1 + n_2 - 2}$	
Expected value of sample mean	$E[\bar{X}]$	$E\left[\dfrac{1}{n}\sum\limits_{i=1}^{n} x_1\right] = \mu$	
Variance of sample mean	$\text{Var}[\bar{X}]$	$E\left[(\bar{X} - \mu)^2\right] = \dfrac{\sigma^2}{n}$	

Note: f_i is the frequency of x_i.

Some Useful Formulae

Description	Symbol	Definition	Hand Computational Form
Regression Correlation Coefficient of n sample pairs (x_i, y_i)	r	$\dfrac{Cov(x,y)}{s_x s_y}$	$r = \dfrac{\sum x_i y_i - \dfrac{(\sum x_i)(\sum y_i)}{n}}{\sqrt{\left\{\left[\sum x_i^2 - \dfrac{(\sum x_i)^2}{n}\right]\left[\sum y_i^2 - \dfrac{(\sum y_i)^2}{n}\right]\right\}}}$
Regression Coefficient in straight line regression $Y = a + bx$	b	$\dfrac{Cov(x,y)}{s_x^2}$	$b = \dfrac{\sum x_i y - \dfrac{(\sum x_i)(\sum y_i)}{n}}{\sum x_i^2 - \dfrac{(\sum x_i)^2}{n}}$
Intercept Term in straight line regression $Y = a + bx$	a	$\bar{y} - b\bar{x}$	
Residual Variance in straight line regression	s_e^2	$\dfrac{1}{n-2}\sum_{i=1}^{n}(y_i - a - bx_i)^2$	$\dfrac{n-1}{n-2}\left(s_y^2 - b^2 s_x^2\right)$
RANK CORRELATION Spearman's Rank Correlation Coefficient of n pairs of sample rankings (x_i, y_i) where x_i is the rank of the ith member of one series and y_i is the rank of the corresponding observations in the second series.	r_s	$1 - \dfrac{6\sum_{i=1}^{n} d_i^2}{n(n^2 - 1)}$ where $d_i = x_i - y_i$	
Kendall's Rank Correlation Coefficient of n pairs of sample rankings (x_i, y_i) where x_i is the rank of the ith member of one series and y_i is the rank of the corresponding observations in the second series.	r_k	$\dfrac{S}{\dfrac{n}{2}(n-1)}$ For each of the $n(n-1)/2$ possible pairs of rankings (x_i, y_i), (x_j, y_j), $j > i$, assign a score of $+1$ if sign $(x_i - x_j) =$ sign $(y_i - y_j)$ otherwise assign -1. S is the sum of these assigned scores.	

Factorials

$\dbinom{n}{x}$	$\dfrac{n!}{x!(n-x)!}$	$\Gamma(n)$	$(n-1)!$ when n is positive, but not necessarily an integer.
$n!$	$n(n-1)(n-2)\ldots 3.2.1.$		$\Gamma(n) = \int_0^{\infty} x^{n-1} e^{-x}\, dx = (n-1)!$

Some Useful Formulae

DISTRIBUTIONS

Name	Parameters	Probability Function	Mean	Variance
DISCRETE Uniform or Rectangular	k the possible number of values	$f(x;k) = \dfrac{1}{k}$	$\dfrac{k+1}{2}$	$\dfrac{k^2-1}{12}$
Hypergeometric	N population size n sample size M number of items in population labelled 'success'	$P(x) = \dfrac{\binom{M}{x}\binom{N-M}{n-x}}{\binom{N}{n}}$	$n\dfrac{M}{N}$	$n\dfrac{M}{N}\left(1-\dfrac{M}{N}\right)\left(\dfrac{N-n}{N-1}\right)$
Binomial	n number of trials p probability of 'success' at each trial	$P(x) = \binom{n}{x}p^x(1-p)^{n-x}$ $x = 0,1,\ldots,n$	np	$np\,(1-p)$
Negative Binomial	r number of successes required x number of the trial when the rth 'success' occurs; alternatively, if c is the number of 'failures' preceding the rth success $q = 1-p$	$P(x) = \binom{x-1}{r-1}p^r(1-p)^{x-r}$ $x = r, r+1, \ldots \infty$ $p(c) = \binom{c+r-1}{c}p^r q^c$ the general term of the binomial expansion of $p^r(1-q)^{-r}$	$\dfrac{r}{p}$ $\dfrac{r(1-p)}{p}$	$\dfrac{r(1-p)}{p^2}$ $\dfrac{r(1-p)}{p^2}$
Geometric	p probability of 'success' at each trial	$P(x) = p(1-p)^{x-1}$ $x = 1, 2, \ldots$	$\dfrac{1}{p}$	$\dfrac{1-p}{p^2}$
Poisson	m average number of random events in a given interval	$P(x) = \dfrac{e^{-m}m^x}{x!}$ $x = 0, 1, \ldots$	m	m
CONTINUOUS Rectangular	α minimum value β maximum value	$f(x;\alpha,\beta) = \dfrac{1}{\beta-\alpha}$ $\alpha \le x \le \beta$	$\dfrac{\beta+\alpha}{2}$	$\dfrac{(\beta-\alpha)^2}{12}$
Negative Exponential	m average number of random events in a given interval	$f(x;m) = me^{-mx}$ $x \ge 0$	$\dfrac{1}{m}$	$\dfrac{1}{m^2}$
Normal	μ mean σ^2 variance	$f(x;\mu,\sigma^2) =$ $\dfrac{1}{\sigma\sqrt{2\pi}}e^{-\frac{1}{2}\left(\frac{x-\mu}{\sigma}\right)^2}$ $-\infty < x < \infty$	μ	σ^2